TECHNOCRATIC
VISIONS

TECHNOCRATIC
ENGINEERS, TECHNOLOGY, AND SOCIETY IN MEXICO
VISIONS

EDITED BY J. JUSTIN CASTRO AND JAMES A. GARZA

UNIVERSITY OF PITTSBURGH PRESS

Published by the University of Pittsburgh Press, Pittsburgh, Pa., 15260
Copyright © 2022, University of Pittsburgh Press
All rights reserved
Manufactured in the United States of America
Printed on acid-free paper
10 9 8 7 6 5 4 3 2 1

Cataloging-in-Publication data is available from the Library of Congress

ISBN 13: 978-0-8229-4748-6
ISBN 10: 0-8229-4748-X

Jacket art: Engineer illustration by alashi on iStockphoto.com, Mexico Skyline
from freepik.com
Jacket design: Melissa Dias-Mandoly

CONTENTS

TECHNOCRATIC
VISIONS

ENGINEERING AND TECHNOCRATIC VISIONS IN MEXICO

J. JUSTIN CASTRO

This book is about experts, technology, and networks in Mexico. It focuses on the period between the mid-nineteenth century and mid-twentieth century, an era some historians align with the Second Industrial Revolution and Mexico's transition to a more industrialized, fossil fuel–geared economy.[1] Some of this volume's authors, however, delve as far back as the pre-Colombian era and other authors bring their analysis up to the twenty-first century. Most the contributors focus on engineers, especially civil engineers, but also on mining engineers, military engineers, architects, and various other infrastructural and mechanical technicians. Bringing together Mexican and US historians with a range of overlapping interests and fields of study, this volume provides a diverse representation of histories about expertise, technocracy, and technology in Mexico.

We examine these engineers and other experts because we think their lives provide important and unique ways of studying how technology and people intersect to create societal and environmental change.[2] Mexican engineers, like their counterparts in other parts of the world, have served as brokers. They mediate between local communities, multiple levels of governance, and multinational and domestic corporations.[3] Their lives reveal a Mexico with competing and intertwining technological vernaculars and lived experiences. We argue that understanding more about the context and use of technological networks, that is, the people who first constructed them, maintained them, and resisted them, is important to creating more informed conversations today about why, how, and by whom these networks are built and to whose benefit.

ENGINEERS AND TECHNOCRATS

So, who gets to count as an engineer? After all, anybody who designs or builds something technically engineers it. In that sense people have been engineering things since they learned to light fires, turn plants and animals into pigments for cave paintings, and knap rocks into tools. The term "engineer" was not even commonly used in Western societies outside of people who worked on war equipment until the eighteenth century.[4] Before then construction works and mechanical creations were carried out largely by local craftspeople or city guild members. The people of the Americas, especially those in more technologically complex civilizations, such as the Inca and Aztec, possessed a variety of specialists. The society in and around Tenochtitlán, the Aztec capital centered in the middle of Lake Texcoco, possessed people who designed and built temples, roads, aqueducts, canals, dikes, docks, and *chinampas*.[5] Colonial practices and subsequent Mexican methods competed and merged with a myriad of previously established practices. The engineers who are the subject of this book were often professionally trained in formal, Western-style architecture and engineering programs in the late 1800s and early 1900s in Mexico, Europe, or the United States. They were part of the broader globalization of the profession. But these professionals interacted with other people who engineered their environments with or without engineers.

The rise of professional engineering programs brought significant changes to Mexico as it did elsewhere. They facilitated intellectual exchanges with other nations even while in Mexico they also networked people interested in strengthening Mexican sovereignty. Confirmed by state, business, and military elites, engineers became determining agents in matters of infrastructural development and certain extraction enterprises and industrial projects, at least in the eyes of those elites and engineers. In the process the elevation of professional engineers attacked the legitimacy of local knowledge and technicians who did not conform to modern methods of standardization, building practices, and concepts of hygiene, safety, and order.

The participation of engineers and other technical experts in government circles in the late 1800s coincided with the rise of technocrats in the Mexican state. As such, *Technocratic Visions* is also about technocracy. Many people who study Mexico use the term "technocrat" as a reference to economic experts who became members of the Mexican government beginning in the 1960s and who became especially prominent in the 1980s and 1990s.[6] Several top administrators and even presidents, including Miguel de la Madrid (1982–1988) and Carlos Salinas de Gortari (1988–1994), trained in economics at Harvard and other prestigious US institutions. Political scientist Roderic Ai Camp called these late twentieth-century *técnicos* "Mexico's Mandarins."[7] Sociologist Sarah Babb referred to them as "new technocrats," economists who were often foreign-trained in "mainstream neoclassical economics [which] disposed them to look favorably on the dismantling of the developmentalist state."[8] In other words, many Latin Americanists associate "technocrats" with these neoliberal economists-turned-politicians.

This concept of "technocrat" evolved from broader visions. Pulling from Karl Marx, Max Weber, and German critical theorists, scholars in the 1960s, such as historian Theodore Roszak, painted technocrats in broad strokes as a (hyper)rationalist network of experts who had risen into the ranks of political power in countries around the world. This coincided with growing critiques of technological development in general. Intellectual Lewis Mumford had decades before lamented the machine-like regimentation of society and the dangers of blind production and consumerism. He amplified these fears in his publications from the 1950s to 1970s, casting aside much of his original optimism about technological possibilities, including about the people involved in developing and implementing them. He was joined in this anti-technocratic crusade by other influential scholars, including Herbert Marcuse, Jacques Ellul, and Roszak.[9] These authors and those influenced by them criticized technocrats as authoritarian, or at least a danger to democracy, because of the very claim made by many techno-bureaucrats that their work was scientific, in part universal, apolitical, and necessary. Some critics feared a complete technocratic takeover of human societies even if at the time it was a "tendency, and not a completely actualized phenomenon."[10] Roszak summed up this conception of technocracy well, defining it as

> that society in which those who govern justify themselves by appeal to technical experts who, in turn, justify themselves by appeal to scientific forms of knowledge. And beyond the authority of science, there is no appeal. Understood in these terms, as the mature product of technological progress and the scientific ethos, the technocracy easily eludes all traditional political categories. Indeed, it is characteristic of the technocracy to render itself invisible. . . . While daily political argument continues within and between the capitalist and collectivist societies of the world, the technocracy increases and consolidates its power in both as a transpolitical phenomenon following the dictates of industrial efficiency, rationality, and necessity.[11]

Roszak then went on to compare societies' technocratic acquiescence to baseball fans and umpires, where the fans are rooting for the teams while the umpire is often the least visible person on the field. But ultimately it is the umpire who judges and enforces the rules of the game.

Other perspectives on technocracy that build on these intellectual predecessors come from more contemporary political theorists who discuss state power, legibility, democracy, and technocracy, such as James C. Scott, who criticizes "high-modernist ideology," which he sees as a strong "self-confidence about scientific and technical progress, the expansion of production, the growing satisfaction of human needs, the mastery of nature (including human nature), and, above all, the rational design of social order commensurate with the scientific understanding of natural laws."[12] He provides a critique of high-modernist thinking as hubristic, controlling, and oversimplified. Intending to make nature and society legible, planners (often engineers) have designed oversimplified environments that have commonly failed to understand the complexity of the natures and societies they attempt to control. In addition, these agents have too often failed to sufficiently consider and incorporate the knowledge of local peoples. None of *Technocratic Visions'* contributors use Scott's high-modernism terminology. Most of us complicate Scott's sometimes overly homogenous portrayal of the state and his tendency to paint engineers as uncritical implementors of state designs. His studies of the state, legibility, and power nonetheless remain valuable.

More recently, *The Technocratic Challenge to Democracy* (2020), edited by Eri Bertsou and Daniele Caramani, provides a more nuanced approach to technocracy and represents an evolving strain of thought about it among some political scientists and sociologists. In the introduction Caramani paints technocracy as a product of the intertwined but tense relationship between nationalism and industrialization. The former, he argues, leans ideologically democratic, egalitarian, and irrational; the latter tends toward hierarchy and rationalization. Caramani further argues that representative democracy has been an attempt to reconcile the two and that technocrats play important roles in these governments. At one extreme, representative democracies can tend toward the hyper-technocratic and elitist. At the other, they can lean toward the irrational and populist. Technocracy is on a continuum; it "takes various grades."[13] The most radically democratic governments still have some need for expertise and technocrats who are not always democratically selected. And even the most technocratic or otherwise authoritarian governments rely on "some form of popular mobilization and inclusion."[14]

The essays in *Technocratic Visions* reinforce this continuum approach, and they provide glimpses into how technocratic agents transitioned from a positivist, authoritarian dictatorship to the revolutionary and often populist-leaning governments. During this transition technocrats increased, not decreased, in number and influence. Many technocratic engineers who participated in the Mexican Revolution did so believing that reliance on technological expertise would increase during the transition from a dictatorship to a representative democracy, not the other way around.[15] However, despite desires by some politicians, including technocrats, to create a representative democracy, Mexico's own attempts to reconcile revolutionary nationalism and industrialization ultimately resulted in the "soft authoritarianism" of the single-party state (1929–2000).

It was during the Second Industrial Revolution that technocrats first became prominent in Mexico. Sometimes their contributions were made behind the scene and not always democratically, but these experts were far from all-powerful.[16] And those technocrats who participated in the Mexican Revolution and the governments it spawned had to contend with the popular forces the revolution awoke. This volume explores the notion that technocrats have tended to be undemocratic, that they have viewed themselves as answerable to science and not political ideologies, local communities, or voters. It also examines the argument that technocrats and technological development have tended to reinforce inequalities instead of dissolving them, despite their rhetoric of technological fixes. The following chapters show that studying the lives of engineers in Mexico provides insights into the complicated but intertwined ways globalization, energy manipulation, revolution, and nation-state consolidation went hand in hand.[17]

PERIODIZATION AND HISTORICAL CONTEXT

The following chapters contribute to discussions about broad, global trends in professionalization and technological development, but they also demonstrate that studying experts and technologies in Mexico's specific context is useful.[18] Mexico's history—its institutions, its politicians, its diverse human and physical geography—were all important to engineers' and other technocrats' worldviews.

Situated between latitude 32° and 15° N, Mexico's environments vary significantly. Large stretches of arid desert exist across the north. There are high-altitude temperate zones in the center and tropical shrublands and jungles in the far south. The *tierra caliente*, or hot country, also includes tropical lowlands along the Gulf of Mexico, such as in Veracruz and Tabasco. Two mountain chains ride the Pacific and Atlantic coasts: the Sierra Madre Occidental and the Sierra Madre Oriental. Their offshoots squeeze the central plateaus. More mountains crisscross the southern tropical regions. Earthquakes occur frequently. Mexico has an abundance of mineral and biological resources, but there are few rivers navigable by large ships that transverse large cities. Historically, living standards, cultures, and connections between central Mexico and other peoples have varied significantly.

Pre-Spanish and colonial legacies influenced postindependence engineers and technocrats in important ways. When Spanish conquerors and their technical experts created Mexico City out of Tenochtitlán, they built heavily on the established network of communities connected by Aztec routes of trade, tribute, and governance.[19] However, pre-Spanish transportation was not constructed to consider transoceanic empires; overseas resource exportation; "old world" crops, animals, and manufactured goods; or an obsession with mining. Colonial engineers worked to map and obtain resources for the benefit of the Spanish empire and the colonial elite. With great difficulty and mixed success, experts and laborers built roads and port works to connect new settlements to agricultural areas, mining centers, and oceanic trade routes. Subsequent national-era engineers had to contend with this geography and these historical-cultural-environmental legacies as they strove to consolidate and strengthen Mexico as a sovereign nation-state.[20]

Mexico also has a long history of professional engineering that has influenced engineers' identities and work.[21] Colonial Mexico was the site of the first professional engineering school in the Americas. Carlos IV established the Real Seminario de Minas in Mexico City in 1792 to revitalize silver mining operations in Mexico. Earlier, Crown officials under Carlos III established the Academia de San Carlos in 1785. It was the first European-style fine arts academy in the Western Hemisphere. It also taught mechanical arts, including architecture. The majority of colonial students were criollo, that is, Europeans born in Mexico. Regularly denied top administrative positions in New Spain, some criollos sought careers as architects and engineers because they represented some of the most prestigious positions they could hold. Like their modern counterparts, these engineers often served as mediators. They regularly served elite interests but often took part in hashing out conflicts between Spanish officials, other criollos, and Indigenous, African, and *casta* communities.[22] As managers and mappers of Mexican people and resources, they also contributed to the establishment of a sense of early national identity by creating discussions about what resources were "Mexican."[23] Mining engineers, alongside a strong cadre of military engineers, were deeply intertwined with the imperial state, beginning a close association between engineers and the government. This tie continued with mining, civil, and military engineering well into the twentieth century. It still exists to some extent, though this relationship has changed significantly and has been challenged by neoliberal policies and the rise of technical schools, such as the Instituto Tecnológico y Estudios Superiors de Monterrey (1943), which have focused on training students to work in the private industrial sector for domestic and foreign corporations.[24]

Other historical realities that influenced Mexico's technical experts were Mexico's political elites' struggle to establish a stable state while fending off foreign invasions. Internal conflicts plagued Mexico's first decades following its independence in 1821. The secession of Texas in 1835 ignited a war between the secessionists and Mexico. Other states attempted secession as well, including the temporarily successful establishment of the Republic of Yucatán (1841–1847). The US annexation of Texas in December 1845 and disputes over Texas's boundaries led to the Mexican-American War (1846–1848) in which the US military invaded Mexico, occupied Mexico City, and took nearly half of Mexico's territory. Fourteen years after the conclusion of the war, France's Napoleon III ordered an invasion of Mexico. French forces eventually took Mexico City and established the Second Mexican Empire, which was ruled by Hapsburg duke Maximilian I (1864–1867). But Maximilian's government and military could never completely suppress the forces of Benito Juárez (1858–1872), who still claimed the mantle of Mexican sovereignty and waged war against Maximilian. Following the end of the US Civil War (1860–1865) and a decision by Napoleon III to remove most of his soldiers from Mexico, Juárez's forces prevailed, executing Maximilian I and more fully regaining control of the nation-state. The US military later invaded Mexico during the Mexican Revolution, temporarily occupying Veracruz (1914) and carrying out the Punitive Expedition into Chihuahua (1916).

These affronts on Mexican sovereignty fueled a strong sense of nationalism among many of the country's engineers who graduated in the late 1800s and early 1900s. At the same time they increasingly worked with engineers from the United States and other industrial nations. The governments of Juárez and especially Porfirio Díaz (1876–1880, 1884–1911) expanded engineering and technical education, creating national preparatory schools, new scholarships, and increased access for nonwhite and poorer students, even while entertaining new strains of racism. Educational leaders reorganized categories of engineering. They separated civil, mechanical, geographical, topographical, and mining engineering. Díaz and his close planners—many of whom were known as *científicos* because of their positivist prescriptions and claims to scientific expertise—realized that they needed more engineers and technicians to carry out the grand infrastructural goals and public works campaigns they believed essential to their motto "Order and Progress." Engineering students, especially civil engineers who sought to direct large public works projects, were infused with the idea that their work was a national duty, that it was necessary for the salvation and improvement of the country.[25]

One complicated challenge that students from the Escuela Nacional de Ingenieros (National School of Engineers) faced was that Mexico possessed immense cultural diversity and limited national infrastructure. Mexico was (and remains) home to hundreds of different Indigenous groups. Today this diversity is often celebrated, but for late nineteenth- and early twentieth-century national political leaders and civil engineers, it was often something to overcome. How could a mishmash of poorly connected, "uneducated" groups of people who spoke different languages protect Mexican sovereignty or build a strong domestic economy in the face of the United States? To defend that sovereignty, political leaders and engineers alike insisted that Mexico had to join the modern world and that its people needed to develop a stronger sense of common belonging, purpose, and exchange. That survival required new methods of repression and inclusion. It required fossil fuels. It required education, infrastructure, standardization, technical expertise, and engineering.[26]

But not everyone was on the same page. Many communities had mixed reactions to projects that challenged local lifeways and worldviews. Not all people within the confines of the political map of Mexico cared about being Mexican; their identities stood elsewhere. Some people had little desire to participate in capitalism or industrialization. Others in these same communities sought out advantages in the changing order.

To obtain greater control, political leaders worked to adapt practices used by Western imperial powers to unite and control their territories. These practices often required new energy sources, technologies, and technical expertise. Through using material infrastructure and public works to help consolidate control, Mexican leaders sought to display strength, a flavored conformity to modern norms, and to rule over their own sovereign nation, even while to many Indigenous groups Mexico was the imperial power. As for engineers, especially civil engineers, they commonly saw themselves as the people to unite, build, save, and "civilize" the nation.

Yet Díaz, late into his more than thirty years as president, often chose foreign experts for the most important projects, which disgruntled the youngest generation of Mexican engineers. This was in part due to a lingering bias that foreigners from the United States and parts of Europe were better and more practically trained than their Mexican counterparts. But it was not solely this rationale. Foreign firms also came with capital to offset government costs and brought new equipment. The heads of these firms often wanted the enterprises they invested in led by their own engineers. Many Mexican engineers nonetheless proved important to these projects: they knew the terrain. Some of them quickly obtained leadership roles. Others served as middle managers, working as go-betweens for top political leaders, business representatives, foreign engineers, and domestic technicians and laborers. During Díaz's final years in office, engineers increasingly gained high government positions and supervised significant undertakings, including a massive potable water project in Mexico City carried out between 1908 and 1910. Many of the most prominent civil engineers to participate in the revolution had worked on Porfirian drainage and drinking water projects. Engineers also worked in the private sector on projects as diverse as wind pumps and electric piano players. Still, many young engineers had become frustrated at this use of foreign experts, limited state funding, and an aging generation of Porfirian politicos who restricted access to the high rungs of government bureaucracies. Students at the National School of Engineering created the Engineering Club in 1908. Its members focused on "national problems." They demanded further "Mexicanization" of the railroads, communications, and postal operations. Most of its members joined the opposition to Díaz in the 1910 election, and they subsequently sided against him during the revolution that followed.[27]

The revolutionary era (1910–1946) prompted new technical needs and calls for further Mexicanization, that is, a renewed emphasis on cultural unity, development carried out by Mexican laborers and experts, and increased Mexican control over industry and infrastructure.[28] And though many soldiers fought in the revolution for personal reasons, the return of lands, and to correct what they saw as previous wrongs, the leadership of the revolutionary faction that won prided itself on being harbingers of a more vibrant modernity. They also continued to rely on international exchanges in their attempts to implement changes. These trends were carried through the revolutionary governments of Venustiano Carranza (1917–1920), Álvaro Obregon (1920–1924), and those political leaders and aligned technocrats who subsequently built the single-party state that dominated Mexican national politics until 2000.

Mexico's proximity to and relationship with the United States also make it unique. The United States grew into a global power during this period, and it did so on lands it took from Mexico. US engineering programs grew dramatically, and US engineers spread their influence into Mexico and beyond.[29] Because of their connected history and geography, substantial changes in one place affected the other, and this became more pronounced as the United States became more powerful. US interests poured into mining, railroads, and the exportation of manufactured goods to Mexico. They greatly influenced, though did not determine, Díaz's rise to power and the subsequent Mexican Revolution. Today the countries are more entangled than ever, participating in a connected infrastructure and a multitude of international and bilateral agreements. Mexico was in 2021 the world's fifteenth largest economy and the United States' biggest trading partner. The long history and politics connected to the abutting and shared space of both countries have at times fueled defensive, nationalistic posturing and at other times intense mutual collaboration.[30]

While some aspects of Western standardization became increasingly common in Mexico by the mid-twentieth century—industrial components and production, scientific equipment, blue jeans and T-shirts—standardization was far from complete. And when engineers worked on projects based on foreign models and technologies, they had to contend with local peoples who resisted them or who wanted to influence the outcomes, which they often did. And even though most architects, engineers, and other professionals worked within international networks that promoted scientific and technical standardization, they had to adapt to Mexico's political and economic conditions, colonial legacies, geography, and cultural diversity. The resulting entanglement created its own distinctiveness in the resulting hybridity. Many state leaders, business owners, and designers embraced aesthetic and symbolic elements that reinforced perceptions of a Mexican uniqueness while simplifying Mexico's complexity—images that could then be projected globally to show Mexico as a sovereign nation that also exhibited a modern conformity to Western standards, systems, and tools.

The technological systems that Mexicans and foreigners in Mexico established during the Porfirian and revolutionary periods dramatically and unevenly altered Mexican communities in ways that are still prevalent. Technologies have a way of reinforcing the politics and power of their creators while also displaying the contestations over them. The technological systems that people developed in Mexico from the mid-nineteenth to the mid-twentieth centuries were created during a period of immense upheaval and change, both domestically and globally. For a variety of reasons, people established technological systems that were often impressive but rarely created in an all-inclusive or equitable fashion. They often came with unintended environmental, economic, and social consequences. Communities and individuals regularly had ideas other than those proposed about how these systems would be used.[31] Nonetheless, many of these systems became well established: roads, sewerage, port works, media, and telecommunications. Their presence has become normalized for many people even as access to them remains inequitable. Too rarely are they historically and critically examined.

If people desire to change these and other similar technological networks—and the social, political, and environmental issues built into them—it will be helpful to have a better understanding of why and how these systems were originally constructed, collaborated on, and fought over.

INTERSECTING HISTORIOGRAPHIES

The authors in this collection pull from a wide variety of primary and secondary sources. Of the latter you will see influences from Mexican public history. You will also find citations of empirical and theoretical works from science and technology studies (STS) scholars, predominately from Mexico, the United States, and Europe. The contributors interact with a number of academic discussions about power dynamics, public space, technology and the environment, and how people determine technologies and how we are shaped by them in return.[32] Many of us engage with the literature on technocrats and technocracy. We also build on social, political, economic, and environmental histories of Mexico.

Mexican scholars have been creating histories about technology, engineers, and other experts in Mexico for some time; engineers have also been writing about and to themselves. Since the late 1800s engineers and architects participated in important journals, including the *Anales de la Asociación de Ingenieros y Arquitectos de México*, *El Minero Mexicano*, *El Arte y la Ciencia*, *Revista Mexicana de Ingeniería y Arquitectura*, and the *Memorias de la Sociedad Científica de Antonio Alzate*. These and several other periodicals were important channels of information exchange among Mexican engineers and architects. They remain important historical sources.

The history of science and technology as a field began in Mexico by the end of the 1960s with the founding of the journal *Anales de la Sociedad Mexicana de Historia de la Ciencia y de la Tecnología* in 1969. The discipline expanded further in the early 1980s, particularly following the establishment of the Sociedad Latinoamericana de Historia de las Ciencias y Tecnología in the city of Puebla in 1982. Juan José Saldaña and other contributors founded the journal *Quipu: Revista Latinoamericana de Historia de las Ciencias y la Tecnología*, which brought together scholars from across the Americas and beyond who sought to document, analyze, and historically contextualize the development of science and technology across Latin America.[33] This momentum drew on STS organizations established in the United States but also on Spanish historians who had begun to argue that the Spanish empire and its American colonies played a much larger and more dynamic role in the development of scientific thought in the sixteenth through eighteenth centuries than had been recognized.[34]

Many of *Quipu*'s first articles also focused on science in colonial Latin America, though the journal went on to publish on a myriad of topics spanning the precolonial to the twentieth-first century. In general, the more analytical articles in *Quipu* frame the history of science and technology within dependency and center–periphery/world systems theorizations while emphasizing unique accomplishments and adaptations in Latin America and the multidirectionality of scientific ideas and technological development in general. Most stridently, Saldaña, the organization, and the journal emphasized that any attempt to understand the networks of ideas and machines that make up the global system of science and technology is incomplete without including Latin America.[35]

Professionalization is a central part of the social and political history of technology and expertise, and a number of Mexican authors have focused on this topic.[36] The same year as the Puebla conference, Francisco Arce Gurza, Mílanda Bazant, Anne Staples, and Dorothy Tanek de Estrada published *Historia de las profesiones en México*, a foundational though largely descriptive book on the rise of academic, medical, and technical professions in Mexico.[37] Of particular importance to many of the contributors to our edited volume have been historical works on engineering.[38] Bazant, for example, published multiple works on the history of engineering education in Mexico. She provided some of the first accounts of engineering programs, especially from the late colonial period until the end of the Porfirian era. She argues that the Porfirian government and engineers made significant advancements in the quality and diversity of engineering programs. This expansion owed much to the increased social stability, foreign investment, and new material infrastructures of the era. In turn, the growth of engineering programs sparked further development. She additionally highlights the importance of foreign education to many of these engineers, who obtained study-abroad experiences as a part of their education, including at the École Centrale des Arts et Manufactures in Paris and at Harvard, Princeton, Columbia, and MIT in the United States.[39] Most recently Mexican historians have focused on the rise of Mexican civil engineering and its ties to architecture, something that some of the contributors to this volume build on.[40]

Historian Raúl Domínguez Martínez's *La ingeniería civil en México, 1900–1940: Análisis histórico de los factores de su desarrollo* (2013) is the most extensive history about Mexican civil engineering during the late Porfirian era and the revolution.[41] Domínguez contends that it's impossible to "clearly explain modern history in Mexico without attending to the history of infrastructure, and that this cannot be explained without reference to the evolution of civil engineering."[42] He emphasizes the important role of the state in the development of Mexico's civil engineering programs by arguing that political leaders saw the discipline as a practical science that was tied to very real needs. While Domínguez states that Mexico faced multifaceted problems bound within historical dependencies on foreign powers, geographic and cultural diversity, class divisions, and a lack of capital accumulation, he ultimately praises Mexican civil engineering as the scientific discipline that did more than any other to counter these deficiencies. All of the authors in *Technocratic Visions* agree with Bazant and Domínguez Martínez that the advancement of engineering in Mexico led to important material and societal changes, but we differ (among ourselves and with other publications) about the ultimate benefits and costs of those shifts.

Mexican social histories on health, public works, and infrastructure have also been particularly influential. A good example is Priscilla Connol-

ly's *El contratista de Don Porfirio: Obras públicas, deuda y desarrollo desigual*
(1997), which contextualized Mexican histories of development and economics during the Porfirian era within a broader discussion of adaptions
between the Mexican state, local labor traditions, and contracting firms.[43]
As the title of her book states, and much like social histories of infrastructural technologies and Mexico since, it is also a work about unequal development. Most of Connolly's book focuses on the massive Mexico City
drainage project and Veracruz port improvements that occurred during the
first years of the twentieth century, both contracted to British financier Sir
Weetman Pearson (titled Lord Cowdray after 1910). Similarly, a number of
other works published in Mexico that have highlighted the relationship between private-state partnerships and the social processes and consequences
of development in Mexico have focused on water networks and use, an
important facet of Mexican life.[44]

Organized STS studies began in the United States shortly before they
did in Mexico. Foreshadowed by prominent intellectuals of the early twentieth century such as Theodore Veblen and Mumford, STS studies commenced with the founding of the Society for the History of Technology
(SHOT).[45] The origins of the organization started with the Humanistic-
Social Research Project (1953–1955), a Carnegie Corporation of New York–
funded undertaking by the American Society of Engineering Education.
Spearheaded by historian Melvin Kranzberg, the networks that built on the
Humanistic-Social Research Project evolved by 1958 into SHOT, which
began publishing the journal *Technology & Culture* the following year.[46]

Historians of Mexico who write about technology have increasingly incorporated the works of STS scholars. There has been a clear influence from
sociologists and philosophers who study the social construction of technologies, such as Wiebe E. Bijker and Thomas P. Hughes, and from Bruno
Latour and his proponents who take the somewhat oppositional approach
of tracing how associations among people, technology, and their environments more broadly construct society.[47] Although the chapters in this volume are rarely in direct conversation with Latour's Action-Network-Theory
(ANT), they tend to reinforce his approach by providing empirical studies
that explore the construction of society through the lens of engineers and
other technical experts who mediated new technologies, national and international political forces, and local concerns.

Recent generations of STS scholars writing about Latin America have
done much to bring Latin America into broader STS discussions. Edin Medina "uses the history of science and technology as a way to understand
processes of political change."[48] She is best known for her book on the intersection of Chilean cybernetics, technological development, the global
exchange of ideas, and the socialist aspirations of Chile's ill-fated president

Salvador Allende—*Cybernetic Revolutionaries: Technology and Politics in Allende's Chile* (2011). *Beyond Imported Magic: Essays on Science, Technology and Society in Latin America* (2014), a collection of essays she contributed to and coedited, reinforces the importance of Latin America in the development, adaptation, and multidirectionality of scientific and technological tools and networks. María M. Portuondo has authored books on Spanish and colonial Latin American science, especially during the sixteenth century. STS scholar Juan C. Lucerna's work on Mexican engineers and their identities is most directly linked to the studies in this volume. Contributors problematize but often reinforce the arguments he makes in "*De Criollos a Mexicanos*: Engineers' Identity and the Construction of Mexico" about the rising nationalism among engineers and the strong connection between engineers and the state from the late colonial period until the mid-twentieth century.[49] Yet, much more work still needs to be done on Mexican engineers who worked predominately in the private sector.

Most US historians currently writing about the history of technology in Mexico were trained in "traditional" history programs.[50] Their publications have largely been political, economic, environmental, labor, or cultural narratives that have placed individuals and communities at the forefront. They have introduced and contributed important discussions about identity, gender, inequality, authoritarianism, and democracy, among other topics, and provide strong historical contextualization and empirical archival evidence that is sometimes found wanting in philosophical and sociological studies.[51] These histories have nonetheless shown perspectives and arguments that reinforce STS social constructionist, co-constructivist, or ANT theorizations.[52]

The writings of economic historians, whose important works long dominated conversations about technological development in Mexico and Latin America, also highlight inequalities but focus more on international commodity chains or the ability of Latin American regions or nation-states to integrate successfully, or not, into the global capitalist market. Whereas many cultural histories of technology have focused on social agency and local or regional histories, many economic works have examined Latin American export economies, Import Substitution Industrialization (ISI), technology transfer, and the integration of transportation networks within global systems. Over the past couple of decades, economic historians have eschewed dependency theory, but the theme of dependence is still fairly common.[53] In these histories, colonialist legacies, personalist politics, weak education systems, capitalist development, a lack of nation-state-market integration, and insufficient specialists have limited the ability of Mexican and other Latin American governments and industrialists to fully integrate technological tools and networks originally developed by foreign specialists. Historical and geographic circumstances, deep institutionalized social inequities, and often political shortcomings have made it difficult for people in Latin American countries to mesh and benefit from the global capitalist system, despite significant natural resources in the region.[54]

CHAPTER OVERVIEWS

The chapters in this collection discuss a variety of experts and technologies, yet they also overlap significantly in their chronology and themes. Roughly half of the essays focus on the Porfirian era and the others examine the revolutionary and postrevolutionary eras, though there was significant continuity between the periods in matters of technology and expertise.

The first two chapters focus on architecture and the rise of civil engineering before and during the Porfirian period. In addition to examining specific people and projects, both chapters spend considerable time discussing the education system that trained these men and the construction materials they used. Marcela Saldaña Solis's "Poetry in Stone and Iron: Architect Emilio Dondé Preciat and the Construction of Modern Mexico City" looks at Mexico City's architecture during the late nineteenth century and the first years of the twentieth century through the lens of architect Emilio Dondé. Dondé straddled the shift of certain educational programs from architecture to civil engineering. His life and work show the contestations and conciliations among global, local, and national forces. In "Revelations from Rediscovered Artifacts of the National School of Engineers' Construction Materials Collection," Lucero Morelos Rodríguez and Francisco Omar Escamilla González use the recent rediscovery of portions of this collection to show how these construction materials correlated with the rise of Mexican civil engineering. They demonstrate how Porfirian officials and experts placed a growing emphasis on obtaining and standardizing materials, though the results were far from complete.

The authors in chapters three and four recognize some of the major accomplishments of engineers and other experts during the Porfirian era, but they are more critical of the consequences that major projects had on the lifeways and health of affected communities, especially those who were not members of the small-but-growing urban middle class and the elite. In "Engineering the Porfirian Landscape: Technology and Social Change in the Basin of Mexico, 1890–1911," James Garza focuses on the engineering and technical elements of the Gran Canal del Desagüe—a massive drainage project—and the voices of communities around Lake Texcoco that were affected by the undertaking. The project created substantial economic, ecological, and cultural changes for people in the region, provoking a number of petitions and complaints based on long-standing traditions but also from a desire to be better incorporated into economic markets and modernization, something members of these communities felt the canal had sometimes hampered instead of improved. Rocio Gomez's "The Preoccupation with Safety: Mining Engineers, Education, and Practice in Modern Mexico" examines how mining engineers became increasingly tasked with limiting occupational accidents. She focuses mostly on the northern state of Zacatecas, arguing that safety was often illusory because too many engineers lacked the hands-on mining experience and local knowledge necessary to truly protect miners.

Chapters 5 through 9 emphasize the decades during and after the Mexican Revolution. Juan José Saldaña's "Revolutionary Technoscience: Science, Industry, Education, and the Mexican State, 1910–1946" is a broad overview of the evolution and intersection of science, industry, education, and the Mexican state during the revolutionary and postrevolutionary periods. He argues that the Constitutionalists and their successors focused even more than the Porfirian government on practical education and the use of technology and technological exports to develop Mexico's resources and to solidify the new revolutionary nation-state. For Saldaña the Mexican Revolution resulted in a genuine revolution in science and technocracy as much as, if not more than, a revolution in democracy and social justice.

Chapters 6 through 8 provide more specific studies about technology and engineers in US-Mexican relations and Mexican nation-state construction. My chapter, "Technocratic Diplomacy: Constitutionalist Engineers as Diplomats to the United States," focuses on how the Constitutionalist revolutionary faction led by Venustiano Carranza used technocratic diplomats, often engineers, to gain US support and a diplomatic edge during the Mexican Revolution. The chapter highlights the transnational character of professional and technocratic discourse.

Jayson Porter's and Pete Soland's essays focus on transportation technologies: roads and airplanes, respectively. In "Punitive Engineering and Military Modernization: Reform, Revolution, and Reconstruction in Mexico and the United States, 1916–1924," Porter tracks how the interactions of the US and Mexican militaries during the Punitive Expedition and revolution spurred national state-building projects in their respective countries. Mexican communities sometimes protested these infrastructural developments; yet at other times they collaborated, attempting to influence military, business, and government planners. Porter argues that the US military took lessons from its road-building and mechanization efforts during their Mexican expedition to the Western Front in Europe during World War I, showing that certain "Atlantic Crossings" originated in Mexico.[55]

Soland's "Flying Machines as a Measure of Mexico: National Reconstruction, the Cultural Revolution, and the Maturation of Mexico's National Aviation Program, 1921–1945" is a study of how revolutionary government leaders imbued aviation technologies and policies with socio-cultural symbolism that furthered their political goals for nation-state unification. These aspirations coincided with a drive to project an image of nationalistic modernity built on cultural inclusion by giving Indigenous names to aircraft. But underlying these names remained a modern Western developmentalist agenda that often cast aside many of the people whom the revolution was supposed to serve.

This volume concludes with Matthew Vitz's "A Social History of Urban Expertise: Between Techno-bureaucratic Rule and the Right to the City in Twentieth-Century Mexico." Exploring Mexico City's built environment from the Porfirian era until the 1985 earthquake, the piece contributes a fitting and powerful examination of how technical experts, bureaucrats, urban middle-class residents, and the city's poor disputed and collaborated on the development of their surroundings. Vitz shows that the consistent demands and influence of Mexico City residents challenge the notion that development lay strictly in technocratic hands. He argues that treating expertise as crucial to livable cities but also problematic for democratic participation provides clues to the possibilities and limitations behind constructing more accountable, equal, and sustainable cities.

Together, these essays exemplify many of the complexities, contradictions, and contestations involved in the creation of modern Mexico. The chapters also show that external and internal political and economic pressures to rationalize people and goods increased substantially from the mid-nineteenth to the mid-twentieth centuries. Many Mexican technocrats argued that a more unified Mexico and interconnected world would bring about more security and material wealth and, in turn, a happier and more prosperous people. Mexico, like much of the world, has become a wealthier place. Its engineers and architects have produced amazing works of structural art, have become leaders in geotechnics, and have expanded infrastructures that allow more people access to goods and services. But it is clear that development in Mexico has been hugely inequitable, and that rationalization as a whole has also worked as a form of control, a sort of repression that has limited as much as expanded individual and communal worldviews and possibilities. Material structures have played a large role in this change. Where we move, what we watch and listen to, how we work, and how we identify have been shaped dramatically by the options and limitations of the technological networks we live within. How and by whom these systems have been built are of paramount importance to understanding where we come from and where we are going. Building a more just and sustainable world requires rethinking these networks and engineering a fairer and more democratic process for future development.

CHAPTER 1

POETRY IN STONE AND IRON

The Architect Emilio Dondé Preciat and the Construction of Modern Mexico City

MARCELA SALDAÑA SOLÍS

From our current knowledge of modern architecture in all countries and of the most perfect classical architectures, without a doubt, knowing these kinds of resources, can we not say that we are cultivating the art of poetry in stone.

—Nicolás Mariscal, "El desarrollo de la arquitectura en Méjico," 1900

On a sunny day in 1898, at four in the afternoon in the conference room of the Ministry of Communications, the jury for the competition for the construction of the Palacio Legislativo (Legislative Palace) met to nominate the winners. To everyone's surprise, there was no first place; none of the projects completely satisfied the judges. Complicating things further, President Porfirio Díaz rejected the jury verdict on the second- through fifth-place proposals. The president subsequently reshuffled the awards to fit his own beliefs.

Thus began a controversial episode in the history of Mexico City architecture, a controversy in which Emilio Dondé Preciat, a prominent architect (1849–1905), was front and center. Díaz ultimately chose foreigners for the second to fifth awards: Italians and Americans. However, when the project finally began, it was Dondé and engineer Antonio Anza, both Mexicans, who were chosen to direct the project. They put forth a myriad of thoughtful designs that built on the concepts of the original competition.

However, before a single stone could be placed, another Mexican architect attacked Dondé for lack of transparency, and the whole project ground to a halt again, not to be picked up until years later by a French architect.

The contest is revealing in many ways. Despite the prominence of *científico* advisers, the competition was not exclusively a matter of rational decisions based on merit and science. Social prestige and personal ties to prominent politicians mattered too. Mexico City was producing a small but impressive number of new architects and civil engineers by the end of the nineteenth century, and several of them possessed abilities that were good enough, when combined with sociopolitical contacts, to obtain commissions for some of the most important Mexico City buildings of the late Porfirian era. Yet foreign tastes still dominated. Elites admired the practical US engineering firms from Chicago that designed impressive buildings with industrial iron and steel. They enjoyed the beauty of Italian and French art and architecture, which combined classical with modern elements.

Dondé is representative of this world. He had traveled Europe, and he borrowed heavily from European designs, techniques, and materials. The Mexican government asked him to do so. But he also prided himself on being a member of a rising class of Mexican-trained architects who could adapt foreign modern techniques and materials to Mexican colonial structures and late nineteenth-century social needs, combining both the local and the foreign into a new modern Mexican aesthetic. His life and work exhibit the continuing importance of familial ties within prominent government circles; Dondé's work was meritorious, but his social standing and political relatives opened doors to generous contracts as much as his skills. Dondé's work further shows the important ties among engineering, architecture, and changing notions of space, spectacle, health, and sanitation—the tensions of Mexico City modernity.

Dondé was born in the southeastern city of Campeche on September 2, 1849, to a family of successful merchants. Elite families formed an integral part of public life, and they held important positions in government and business. They created client networks of marriages, ritual kinship (*compadrazgo*), and friendships, and these networks often enhanced business ties and shaped politics. The Dondé Preciat family acted no differently. Rafael Dondé, Emilio's older brother, moved to Mexico City to study law at the Colegio de San Ildefonso, drawing on the family's economic and social resources. He later served in the Mexican Senate and developed business relations with important investors such as Thomas Braniff, Félix Cuevas, and others. In 1865 Emilio followed his brother to Mexico City to study engineering and architecture at the Academia de San Carlos.[1] His brother and his education helped Emilio join circles in which the social elite and the government required new technical experts to turn modern aspirations into the material face of the nation. In this sense Emilio Dondé represents an increased inclusion of technical experts into Mexico City's social and political elite. Architects and engineers gained social status as the people who could make Mexico City more than a "city of letters"; they could create a city of reinforced concrete, iron, and glass that could compare to its European counterparts.[2]

His architecture is also important in and of itself. He was prolific in his construction, and his architectural conservation and renovation works were extensive. His buildings were genuinely innovative and shaped the aesthetic of Mexico City. His influence is still present to this day, and his projects exhibit a global movement of ideas and how in architecture they were adapted to local circumstances. Dondé is representational of elite and middle-class dreams for a new architecture that combined elegance with modernity and showcased prosperity and "Order and Progress."[3] He also exemplifies the rising prominence of technical elites, and his works became synonymous with Mexican urban modernity.

At the turn of the twentieth century many elites and commoner urban residents in Mexico, though by no means all of them, were excited about the potential progress of modernity.[4] This was especially true among engineers and architects who believed themselves to be in no small part the designers of the modern world. In the words of Mexican jurist, academic, and writer Manuel Gustavo Antonio Revilla: "The invention of previously unknown architectural forms corresponds to the modern era, or rather, the adaptation of the old to the new needs of the contemporary society. . . . To the great contemporary architects belong the innovations that constitute the marked progress in the architecture of our days."[5] Revilla's words demonstrate the value and status that many Mexico City intellectuals invested in architects, and also that some of these intellectuals saw the work that architects were doing as influenced by societal demands, by the cultural and market forces unleashed by global industrialization and capitalism. Using industrial iron and reinforced concrete, architects combined material innovation and new conceptions of space, expanding possibilities for new types of buildings and the reformation of old ones, changing how people interacted with each other and objects in those spaces.

OLD SPACES, NEW PROJECTS

The War of Independence (1810–1821) shattered Mexico's bonds with Spain, and devastations brought the colony's economy to a standstill. In the following decades, the changing governments could not maintain order and establish solid institutions. Different political actors—centralists, federalists, liberals, and conservatives—struggled to impose their vision of government, and disorder reigned.

A series of political conflicts in the middle of the nineteenth century undermined the integrity of the nation, including Texas's secessionist movement (1835–1845) in the north and Yucatán's in the south. Although the Yucatán rejoined Mexico in 1848, the Republic of Texas joined the United States in 1845. The United States seized California and New Mexico in the Mexican-American War (1846–1848), and Mexico sold the Mesilla Valley to the United States in 1853. Later, during the French Intervention (1861–1867), the forces of France's Napoleon III invaded Mexico and established a monarchy ruled by Hapsburg duke Maximilian I.[6] Porfirio Díaz gained a reputation as a hero during the war against France. He took the presidency by force in 1876, which he held with one interruption until 1911. Diaz used dictatorial powers to restore peace and stability and enhance economic development.[7]

Industrialization, trade, and foreign investments were key components of Porfirian economic strategy. The científicos and their aligned political leaders hoped that foreign investors would develop mining and timber production and expand livestock and other agriculture that could flourish in the nation's diverse soils and climates. Both foreign and domestic enterprises built railroads that connected the cities, mining hubs, agriculture, and ports, accelerating travel and transcending geographical barriers. Engineers who constructed bridges, roads, and railroads contributed much to the nascent economic expansion.[8]

Mexico City architects and engineers had to consider the area's environmental foundations and colonial legacies. Its precursor, the old Aztec capital of Tenochtitlán, had been founded on spongy ground on an island in Lake Texcoco. After the conquest, colonial buildings went up on the solid foundations of pre-Hispanic temples. Officials worked to drain the lakes. Wide and heavy walls, some made of *tezontle* (very light, porous volcanic stone, dark red in color, used in construction), some with remains of pre-Hispanic temples, marked government and Church buildings.

In the nineteenth century, however, many of the massive structures continued to subside into the soft subsoil. They had become increasingly dilapidated by Dondé's time. Small merchants, artisans, and minor trades workers often occupied these buildings in combined houses and shops. Some of these heavy constructions exist to this day due in no small part to the fact that they remained occupied, and the most important ones have undergone significant conservation procedures. On the outskirts of the city, in places such as those discussed in James Garza's chapter, were tiny houses made of natural products such as wood and adobe.[9] By the mid- to late nineteenth century the widening of streets and inclusion of public works transformed the city.[10] The city had changed its appearance since colonial times, but on the other hand, it retained many colonial buildings. Thus at

Fig. 1.1. "Introduction of drainage in a street." *Source*: Fototeca Nacional INAH, 89905, ca. 1925, Secretaría de Cultura-INAH-Mex, Archivo Casasola Collection, Mexico City. Reproduction authorized by the National Institute of Anthropology and History.

the time Dondé arrived in the capital, we must imagine it as a city in transition from its colonial configuration to one on the verge of a new, Mexican modernity.

Mexico City established new services in the late nineteenth and early twentieth centuries that improved the lives of many inhabitants. An aqueduct supplied residents with drinking water from Xochimilco. A massive drainage system captured wastewaters and channeled them out of the valley. These public works made life easier for many residents, and improved sanitation, relieving measles and scarlet fever epidemics that had affected the city. Despite these developments, social inequality or structural and environmental problems persisted. Garza's chapter in this volume shows that villages outside of Mexico City sometimes suffered less access to goods and services because of the new drainage canals. But regardless of the new drainage system, flooding remained a significant challenge. Engineers, students, and an army of laborers quickly built conduits to benefit elite and middle-class neighborhoods, while some poor communities continued to go without or had to walk long distances for access. Fire hydrants were placed most abundantly in wealthy neighborhoods and commercial areas. The designers of the Xochimilco aqueduct and its associated water projects were unaware that they would cause serious environmental problems by depleting the Xochimilco springs and that the lowering of the area's water table increased further subsidence.[11] Nonetheless water access did improve for many people, while urban streetcar lines replaced some transportation provided by animals and people. Many political leaders, foreign diplomats, experts, and a good part of the general populace welcomed these developments.[12]

Mexico City "positively" transformed its appearance. The city expanded its limits to former haciendas, which had traded in farmland and pastures for large urban residences. For these new homes, the city council of Mexico City offered a tax exemption to include a garden of several meters in front of a property.[13] In addition to gardens, lakes, and greenhouses, city laborers constructed spaces for vehicles—first for horses and then for cars—in the back of many of these homes. Connecting distant points, new roads were created. For example, the Paseo de la Reforma (1864) linked the old Chapultepec park to the city center. These new avenues had sidewalks and public lighting; benches, vases, and statues of prominent figures decorated the streets. Such elements were often markers of a vibrant cultural nationalism in most modern European cities. Mexico City planners saw them in a similar way.

THE ACADEMIC FORMATION OF DONDÉ AND HIS PEERS

In colonial Mexico guilds dominated urban knowledge industries about construction. The guild was a corporate entity that brought together people working in the same activity or profession. Well-established members controlled the dissemination of knowledge, the training of new members, and the provision of raw materials and prices. They possessed significant leeway to control the services they offered. In 1785 the Crown founded the Academia de San Carlos to teach painting, sculpture, architecture, and,

TABLE 1.1. STUDY PLAN FOR CAREER PATH FOR ENGINEERING AND ARCHITECTURE AT THE ACA-
DEMIA DE SAN CARLOS, 1858

1st Year	2nd Year	3rd Year
Trigonometry	Conical Sections	Rational Mechanics
Analytical Geometry	Differential and Integral	Descriptive Geometry
Drawing and Explanation of	Calculus	Composition and
Classical Orders	Monument Styles	Combination
Ornamental Architecture	Inorganic Chemistry	of Building Part specifically
Physics		for Construction
		Elements of Geology
		Minerology and Topography
4th Year	**5th Year**	**6th Year**
Aesthetic Theory of	Applied Mechanics	Construction of Common
Constructions	Construction Theory and	and Iron Roads
Applications of Descriptive	Vault Aesthetics	Construction of Bridges,
Geometry	Composition of Buildings	Channels and Other
Art of Drawing and	Aesthetics of Fine Arts and	Hydraulic Works
Projecting	Architectural History	
Machines	Geodetic Instruments and	
	Their Application	
7th Year	**Certification**	
Practice in Architectural	Preparation of a project and defense before a jury. The	
Engineering	work should present architectural plans and display of	
	calculations	

Source: Israel Katzman, *Arquitectura del siglo XIX en México* (Mexico City: Editorial Trillas, 1993), 53.

more broadly, the mechanical arts. The academy helped spur a republic of
material arts that slowly professionalized industries in ways that challenged
the old guild structure.

With the Academia de San Carlos well established, instructors began
in 1856 to include more formal training in the new fields of architecture
and civil engineering. The idea was to modernize instruction by strengthening coursework through the incorporation of new, practical, hands-on
training outside the classroom in addition to theoretical training. Students
traveled outside the city to experience construction challenges up close and
personal. Over the next decade they traveled to construction sites for bridges, roads, and eventually railways in hopes that these future engineers and
architects would solve problems by executing projects that had to that point
been carried out predominately by foreigners. In other words, the move to
hands-on training and increasing the ranks of Mexican civil engineers and
architects grew substantially during the Díaz era—as shown in the chapters

by Gomez and Morelos and Escamilla—but they began during the governments of the Liberal Reform and Maximilian I.

Table 1.1 shows the classes students took in 1858. The curriculum combined math, physics, art and art history, geology, theory, and construction methods. The last year of training and certification required on-the-ground practice and the development of an exceptional student design with the anticipation that it would be realized. The curriculum was meant to prepare professionals to improve social and infrastructural needs and modernize Mexico's urban image.

These reforms created the foundations of Mexican civil engineering, with the goal that graduates would effectively address the structural and modernization needs of the state. In this sense the country crystallized serious modernization aspirations around architects and early generations of civil engineers who studied at the academy, even though some students also received scholarships to study in Italy, France, and the United States.

In 1867 engineering and architectural graduates from the Academia de San Carlos taught at the newly founded Professional School of Engineering, which became the National Engineering School (ENI) in 1882. They trained the engineering students by borrowing from the two key disciplines taught at the academy, combining technical knowledge on calculating static loads with architectural principles of decoration and classical order. As Morelos and Escamilla discuss, in 1882 the ENI developed an extensive library, a Construction Materials Laboratory, which corroborated the standardization and firmness of construction materials, especially iron, bricks, and reinforced concrete.[14]

When students at the Academy concluded their studies to obtain the title of engineer and architect, they presented a final project. The blueprint of that project was then submitted as a formal project, potentially to be carried out. Dondé and his partner Miguel Pérez proposed redesigning the San Pablo Hospital.[15] It was a complex issue because they had to adapt the existing structure of the building, which had in times past housed a school, to the needs of the current hospital. The building retained training halls but needed more space for patients and employees. Early on Dondé was interested in refurbishing older buildings; he was also interested in health. Modern building design, city planning, and sanitation went hand in hand for him. In this way he was similar to other designers of the era, interested in sanitation, beautification, and new methods and materials for construction. His and his partner's project was accepted by their advisers, who gave favorable opinions of the two, allowing them to graduate in 1870.[16]

Dondé's professors and peers alike considered him an exceptional student. He received awards for his outstanding work at the end of each school year. Some teachers additionally named him a substitute to teach their courses. This started his career as a teacher, an activity that he carried out until the end of his life. But his reputation as an outstanding student alone was not enough to guarantee him a place in the world of top builders.

Without a doubt his knack at engineering was an important part of his success in the field, but strong relationships within elite society and government circles were also important. The government planned to expand public works, and it saw engineers and architects as particularly valuable to these endeavors. The Association of Engineers and Architects of Mexico, founded in 1868, helped bring together a group of peers who also influenced politics. The association sought a space where its members could discuss issues related to their discipline and the nation's governmental, economic, and cultural problems. It was also a place where these professionals could mingle with other elements of Porfirian industry, commerce, and governance.[17]

CONSTRUCTION WITH A TOUCH OF MODERNITY

Mexico increasingly imported machinery and tools from Europe and the United States from 1870 to 1910. The new technology transformed all aspects of urban life. Streetcars began transporting people in greater numbers and with greater speed. Household technologies such as sewing machines became accessible for more people. Modernity was not created once and for all, but it did unfold over time and in different spatial ways.[18]

In the field of construction, modernization largely meant the use of new construction materials. The Second Industrial Revolution from the mid-nineteenth century until the first decades of the twentieth century created a new architectural panorama. With new materials such as industrial iron and reinforced concrete, architects and engineers could design buildings with new shapes and heights. Mexicans had as examples the Library of Saint Genoveva in Paris (1843–1850) and several new train stations, such as the Frankfurt am Main station in Germany (1881) and the Gard du Nord in Paris (1860). They interacted with foreign experts such as US engineer James Eads, who had constructed the Eads Bridge in St. Louis and who proposed to build a railroad to carry ships across the Tehuantepec Isthmus.[19] These projects were inherently intertwined with a grander, more rapid, and interconnected sense of space and movement. Mexico was quickly integrated into this constructive currently spreading across Europe, Australasia, the United States, and other parts of Latin American. A new economic surplus built by Porfirian economic advisers encouraged the construction of buildings that advertised political power and a thriving society.

Dondé had close links with the Díaz government, which allowed him to obtain important commissions, including those connected to world's fairs. He was, for example, involved in the International Exhibition in Paris in 1889. World's fairs were a space to showcase resources, technological advances, and the modernity of nations, promoting trade and communication among the attendees.[20] Some buildings for the international exhibitions were samples of industrial progress because the countries took pains to build them to demonstrate advanced techniques. The Eiffel Tower (1889), the Crystal Palace of Madrid (1887) and the Crystal Palace of London (1851) became famous monuments. Architects also used the exhibitions to meet and discuss challenges regarding the restoration and conservation of monuments. These exchanges provided criteria for the preservation of historic buildings. Dondé participated in government commissions in Europe focused on these types of exchanges and general observations. In the words of the minister of development: "We hope you take advantage of your stay in Europe, and we recommend you study the different construction procedures used in civil and architectural works that are built in Paris, Berlin, Belgium and London, when you return to Mexico to report on the matter."[21]

These words demonstrate the familiarity between Dondé and the Díaz government. They also show the Ministry of Development's deep interest in understanding European construction. The ministry was particularly attracted to the use of industrial iron and steel. Mexico did not obtain an iron and steel mill until after 1900, but the government managed to import the materials for large construction projects.[22] Europe and the United States used industrial iron, steel, and reinforced concrete more frequently; Mexico followed soon thereafter. Industrial iron became prominent in Mexico City and other major urban areas during the last decades of the nineteenth century, in no small part thanks to its export from Berlin, Paris, and Philadelphia.[23] During the first years of the 1890s, the Porfirian government raised tariffs on certain manufactured goods but lowered them on resources for industry and construction.[24]

Industrial iron spread in Mexico City, and it was used to construct new buildings and to refit viceregal buildings. In these colonial structures architects and engineers mostly used the material to construct or replace stairs and reinforce roofs. In addition, central patios, which served as access points to different rooms, had been roofed to protect residents from the weather. Lanterns provided light in the stairwells. Designers and construction workers adapted iron to serve these purposes, combining modern materials and taste with colonial architecture.

An example of those modifications is the Dondé brothers' house. Emilio joined two adjoining houses and transformed them into one, even more stately home. He roofed the patios with iron covers and glass. He also designed a new iron-based staircase and skylight to light up the entire home. In addition, he built bathrooms and a new kitchen. Some of these changes can still be seen in the building. In fact, in one part of the home there seems to be an empty elevator shaft, which would be yet another fascinating innovation if verified.[25]

Dondé and other builders altered public space. For example, they framed and electrified shop billboards to attract and impress visitors. Sidewalks for pedestrians lined the city's principal avenues. Women and men strolled leisurely, and children savored ice creams on these beautified avenues. Patrons enjoyed their coffee in restaurants overlooking downtown Mexico City's beautified spaces. A mix of iron, glass, and electricity joined with colonial masonry and floorplans. The builders created a modern Mexico City aesthetic that was unique, even if foreign-influenced.

THE LEGISLATIVE PALACE PROJECT

One of the most controversial and high-profile construction projects in late nineteenth-century Mexico was the Legislative Palace. The palace was to house the senate and the senatorial offices in addition to some other government entities. The Porfirian government planned to inaugurate the building during the centennial of Mexican independence in 1910. It called for a competition for the construction and invited architects in the hopes of attracting international attention in April 1897. The prize for first place was 15,000 silver pesos, which was the equivalent of a new house in a fashionable neighborhood. The winners of second and third place would receive 6,000 pesos. The fourth and fifth places would, respectively, win a gold and a silver medal plus 500 silver pesos.[26] Dondé participated as a judge in the competition.[27]

The jury handed the important prizes to foreign architects, though it could not agree on a winner. Many Mexican elites admired the prestige of designers from Europe and the United States, and the jury and President Díaz did not differ in this regard. For this reason an Italian company and the architects Peter Joseph Weber and Adamo Boari shared second place. Weber and Boari both worked for Burham & Company in Chicago, though Boari was Italian-born. The Díaz administration later hired Boari to create several high-profile buildings in Mexico City, including the Palacio de Bellas Artes (Palace of Fine Arts, 1904). Paolo Quaglia garnered the third-place prize.[28]

Architect Antonio Rivas Mercado attacked the jury's "whims" and mismanagement. He published aggressive opinion pieces in various newspapers to sabotage the selections for the Legislative Palace construction. Some journalists chimed in, adding a nationalist rebuke in their critiques. While they never mentioned the foreign origins of the winners, they agreed with Rivas Mercado that the awardees did not design the best projects. Praising the work of Rivas Mercado, these writers insinuated they wanted a local champion.[29] Rivas Mercado himself had obtained complete plans

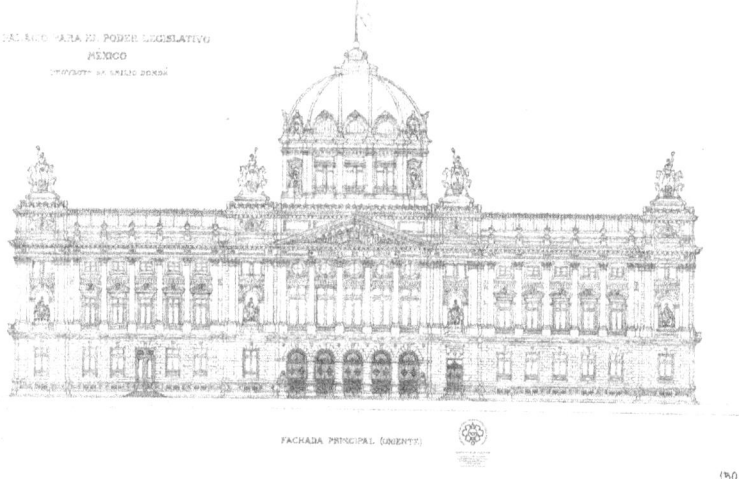

Fig. 1.2. "Palacio para el Poder Legislativo, México. Proyecto de Emilio Dondé," *Source*: Palacio Legislativo, plane 150, Archivo Histórico Jorge Enciso, CNMH-INAH, Secretaría de Cultura-INAH-Mex-, Mexico City. Reproduction authorized by the National Institute of Anthropology and History.

from the Secretary of Communications and Public Works (SCOP) and must have passed the documents to the newspapers.[30] This ruckus shows some divisions within Porfirian social circles; Rivas Mercado, several journalists, and at least one SCOP insider were clearly on friendly terms.

In response the government called for a new design for the Legislative Palace, blending international and Mexican elements. Possibly influenced by public desires, Díaz this time looked for Mexican architects for this hallmark building. He tapped Dondé and the engineer Antonio M. Anza for the task. With Dondé and Anza at the helm, things moved more quickly. They originally based their work on Quaglia's design but quickly veered from the script and drafted 158 detailed plans for alternative facades.[31] Dondé also looked for inspiration abroad. In 1899 he traveled to Europe and the United States to study various parliamentary and congressional buildings—for example, the new Reichstag (parliament) in Berlin—and to inspire his own project. In fact, his proposal bears a striking resemblance to the Reichstag. In 1901 he made another trip to Europe to purchase material and machinery, such as heavy building cranes. He probably also ordered iron abroad.[32]

Yet Rivas Mercado continued his withering criticisms, homing in on Dondé and Anza. Rivas Mercado chided the architects for drafting a childish and poorly executed design, though he never dared to criticize Díaz for assigning the work to them. Neither did Rivas Mercado mention that the jury had rejected his own project, probably to cloak his self-interest. Dondé refused to respond to the accusations, but an uproar ensued. Dondé finally had to resign his commission; Díaz and his government yielded to published opinion and canceled the project.[33]

It was not until 1910 that construction on the Legislative Palace resumed. Instead of showcasing a complete building, the first stone was laid at the celebration of the centennial of Mexican independence. It was not Dondé or Anza at the head of the renewed project, but the French architect Émile Bénard. The building began with great excitement. A powerful metal skeleton went up. Beautiful photographs of the construction process remain, though bad luck haunted the building. By the outbreak of the Mexican Revolution, the palace was still unfinished.

This episode yields insights into the intersection of the Mexican state and the creators of buildings and public works. Guidelines were ignored, public opinion proved important, and Díaz ultimately made the final call. This dependence on favoritism was the opposite of a rational, technocratic process. Even if Díaz surrounded himself with important "scientific" advisers, he nonetheless usually wielded power enough to override them; he could definitely reject a jury decision that was tasked with awarding a building contract. Personalist politics in many ways remained supreme. That could well be advantageous for people like Dondé, who came from wealthy families with relatives in the top political circles (at least until a public controversy ensued). But it shows that the building modern buildings was not some "modern" impersonal process determined only by experts.

This process also displays the intersection of foreign designers and firms, domestic experts, and the Mexican government. The Porfirian government took an active role in Mexico City building projects and the hiring of foreign and domestic firms. Government officials drove the educational changes that encouraged the growth in architectural and engineering programs while also funding students and graduates to travel abroad to report on foreign developments and to obtain materials. Despite local innovations in the adaptation of foreign styles to historic Mexican buildings, the Porfirian elite still often hired foreigners for the most dramatic projects. Personalist, and even popular, politics remained important.

COMMERCIAL BUILDINGS

In addition to legislative buildings, modern buildings that housed department stores transformed the Porfirian city of the late nineteenth century. Members of the small but growing middle class bought a variety of national and foreign merchandise. The wealthy spent lavishly on European clothes, and even the poor could often afford cigarettes produced on French rolling machines that the manufacturer El Buen Tono used.[34] Many of these foreign businesses—and local ones that sold foreign goods—constructed new, beautiful, and enticing stores to increase the chances that their investments would pay off.

The iron for the structures and for the glass roofs gave a different appearance to the buildings; designers incorporated more natural light in merchandise displays, using it to create dramatic scenes in new glass showcases. French styles dictated the guidelines for most upper-end department stores. French businesses had taken a leading role in luxury textile imports in Mexico City. Lightweight staircases formed a central part of many of these buildings, and they were illuminated by sunlight that passed through glass roofs. In Mexico buildings built with iron skeletons and glass covers have been preserved and exist to this day. For example, the El Centro Mercantil was inaugurated in 1899, and El Palacio de Hierro was completed in 1891, which both remain prominent landmarks.[35] One French journalist wrote in 1904 that shopping in the Palacio de Hierro was like being in Paris itself, particularly referring to the change in space and lighting. If a French visitor had been there "thirteen years before (1891) . . . there were small shops without air or light. . . . where clients in semi-darkness spent two hours to buy the article the desired."[36]

Other businesses followed suit as they attempted to survive amid fierce competition. They took advantage of the old colonial spaces and placed attractive covers and stairs in central patios. With these renovations, stores including Sorpresa y Primavera Unidas—a clothing store located a short distance from the Mexico City Zócalo—presented a modern image of themselves and their products for their clients.[37] Fourcade and Company fused two older stores into a larger department store. The owners and their hired workers transformed the structure into an elegant three-story building where customers hunted for silk fabrics, wool, cotton, linen, board games, perfumery, and all kinds of bridal gifts and wear. The showcases were elegantly arranged to highlight the eye-catching luxury items sold as necessary for entry into the higher rungs of urban society.[38]

In August 1893 Dondé helped renovate Sorpresa y Primavera Unidas. He designed an elliptical staircase and placed it in the center of the building, which was an arrangement similar to the staircase in the Parisian department store Le Bon Marché. Dondé found in the design of stairs an architectural resource to highlight the beauty of the building. Nonetheless, Sopresa y Primavera Unidas announced bankruptcy in 1904. Incorporating a new, modern aesthetic may have made downtown department stores more competitive, but it did not guarantee success in a crowded market dependent on a limited clientele wealthy enough to purchase expensive items. In 1907 German architects Hugo Donner and Luis Beicmaster bought the building. They further renovated it, expanding the metal structure to its entirety, a feat they astonishingly completed in three months. The building still possesses the same metal framing today.[39]

Another example of commerce-centered construction is the Lagos de Maracaibo chocolate factory.[40] Dondé worked on this building as well. In addition to similar modern trappings, his work on this structure also showcases his interest in sanitation and the influence of changing norms in gendering public spaces. In his plans Dondé included provisions of the Mexico City health code, especially in his design for hydraulic installations, grease traps, and drainage installations.[41] He also assigned four bathrooms, likely signifying the separation of women's and men's restrooms. As women joined the workforce, separate bathrooms were in greater demand. Architects including Dondé were expected to consider and shape hygiene behavior in ways that affected not only class but also gender.

For the factory Dondé designed an ornate facade. It boasted semicircular openings in accordance with the taste of the time and incorporated wrought-iron windows and doors. The staircase was made with an iron frame, and as with his house, he constructed a roof from iron and a glass cover over the staircase to give natural light to the entire place. Late nineteenth-century modern architecture sometimes conjures images of cold metal and the dark and grime associated with industrialization, but many designers, including Dondé, emphasized light and hygiene. And for city planners at the time, the two were intertwined. They and associated health officials promoted open air and light as beneficial to health. The design of the facades and the inclusion of spaces such as a living room, bedrooms, and service rooms indicate that the building not only functioned as a factory, but that it was also home to the business owner and possibly others. The commercial and domestic were combined into one modern, functional space.

Many of the old retail stores fell by the wayside in the 1890s, replaced by large department stores that were built in and around the old city square, sometimes on colonial foundations. They forever altered consumer patterns and architecture. Some of the most prominent department stores in Mexico City today have origins in this period. And for those that did not last, many left buildings that did. A sign of these buildings' importance is the fact that they were the tallest buildings in the city at the time.[42]

MEETING AND RECREATION AREAS

Café Colón was a fine dining restaurant on the Paseo de la Reforma and a meeting place for socialites. Celebrants of all sorts regularly devoured a delicious variety of desserts, cakes, and ice creams. But it was not only the gastronomic offerings that made this place such an attraction. By 1891 the Café advertised its new, innovative electric lighting. Electricity was still a

Fig. 1.3. "Café Colón, facade." *Source*: Fototeca Nacional INAH, 123790, 1935–1940, Secretaría de Cultura-INAH-Mex-, Archivo Casasola Collection, Mexico City. Reproduction authorized by the National Institute of Anthropology and History.

novelty, a spectacle of sorts, and it increased sales, as people could occupy the terrace late into the night. A path to the café was also lit up.[43] The restaurant is an example of the modifications made to buildings thanks to iron, electricity, and engineer Dondé. He incorporated an iron and glass structure to create additional space next to the old building; it was one of the few sites made this way in Mexico City.

The ground floor terrace offered a privileged view of the new Paseo de la Reforma, itself a showcase of global tastes and high society, but also state power and nation-state construction. The Díaz government was in the process of constructing an "official version of history" through monuments along the Paseo, most prominently the Aztec King Cuauhtémoc. The Cuauhtémoc monument possessed a strong nationalist element, building on pre-Hispanic Mexican imagery to show a long history of pride, courage, strength, and patriotism—to adopt it for a modern Mexico. As Pete Soland discusses in his chapter, revolutionary officials would build on this practice, giving Indigenous names to modern machines. The Cuauhtémoc statue was also nationalist in that it was created by Mexican engineers and artists. The monument, however, also displayed a metropolitan sophistication. Its existence was made possible due to training and materials adapted from other parts of the world, and their construction itself was a display of Mexico's technological progress meant for locals, but also foreigners, to witness.[44]

Another building where well-to-do residents of the capital could enjoy spending their free time was the Baños de Factor, which was what we now call a spa. The owners and their employees provided therapeutic and relaxation treatments. The facilities were modern and emphasized comfort, having steam, swimming pools, showers with hot water, and bathrooms. All the facilities were doubled, one for men and the other for women, thus ensuring "order and morals." Again, it is clear how the concept of modern design incorporated moral norms that included greater hygiene and sanitation, and also the separation of genders in many of these spaces.

Dondé and similar designers hoped to increase hygiene practices and their own notions of beauty. In the renovations he made for the company, he designed and oversaw the incorporation of hydraulic installations. He also created the room for the machines and pumps necessary to bring hot and cold water into the tanks. He further created an extensive garden, and inside the halls simulated an art gallery, placing vases, sculptures, mirrors, and furniture from Vienna to provide an environment in which one was "to think of the lavish baths of ancient Rome."[45] The work showed an affinity for ancient decadence but also the influence of modern North Atlantic trends in urban gardens and hygiene. As the writings of engineer Alberto Pani that Justin Castro discusses in his chapter show, these sanitary engineers emphasized increased health, not only in the private sphere but also in spaces for certain segments of the public, such as Los Baños el Factor. These designers equated cleanliness with civility and modernity.

Although Dondé often designed projects that benefited the well-being of the privileged classes, some of his buildings served all social classes, who after all lived and interacted in certain parts of the city, such as the city center. For example, Dondé designed the San Felipe de Jesús church located on Madero Street in downtown Mexico City. Madero Street was (and still is) one of the busiest streets in the city. People of all social stations traveled the street and gathered in the church to worship. Dondé used an innovative iron structure to make the building lighter, the walls thinner, the roof higher, and the structure more cost-effective. As a shared space open to members of all classes, the church, alongside other urban transformations, gradually impacted the quality of life of most residents of the city.

Economic growth and relative political stability during the Porfirian era fueled an increase in Mexico City construction that gradually transformed it into a modern city, though one deeply entwined with its colonial roots, both physically and culturally. During the transition, engineers and architects such as Dondé found in the two cities—the colonial and the modern—the channels to adapt to new societal desires and perceived sanitary needs, at least for those who could afford them. And some of his work, such as the San Felipe de Jesús, catered to a broader swath of urban society. In addition to outside push factors, these changes fueled a substantial demographic increase that accelerated more than anyone could have imagined by the mid-1900s, creating new challenges for the capital's engineers and architects, and for the city's residents as a whole.

Dondé's work exemplifies the new specialized architectural and engineering studies and techniques that later Porfirian and revolutionary professionals built on. He was an exceptional talent. He was also perhaps somewhat typical among Porfirian architects in that he came from a well-connected family and he worked within academic and administrative circles. These connections opened the doors that allowed him to create and renovate the buildings in which he bridged the past to a new global modernity.

Dondé is a clear example of how Mexico became involved in the modernization movements of the late nineteenth century and the first decades of the twentieth century. Cultural transfers had grown alongside steamer ships, railways, and telegraph lines. Mexicans who studied in the United States and Europe returned to Mexico in hopes of making their country more metropolitan and also more secure in its sovereignty. Joining peers in international gatherings, Mexican engineers borrowed heavily from construction and conservation practices in Europe and the United States, but they adapted them to local structures and circumstances. However, there were still too few architects like Dondé to completely wean the Mexican government and social elite from European specialists. They continued to work and influence the shape of Mexico City and other urban areas too. In some ways, their regular inclusion exhibits a dependency, but in others, it also shows a vibrant space of cultural exchange.

The materials that were a part of this exchange were as important as the ideas. Local and foreign architects and engineers began to use new construction products with regularity: cement, industrial iron, steel, and glass. These materials allowed greater resistances within structures, increased construction speed, and lowered costs. They also sparked the rise of Mexico's first iron and steel mills, glass factories, and cement production sites. All of these industries grew in prominence in subsequent decades, and all of these materials remain mainstays of Mexican architecture. The breadth of architectural poetry had expanded; no longer was it just a poetry of stone, but also a poetry of glass and iron.

REVELATIONS FROM REDISCOVERED ARTIFACTS OF THE NATIONAL SCHOOL OF ENGINEERS' CONSTRUCTION MATERIALS COLLECTION

LUCERO MORELOS RODRÍGUEZ AND FRANCISCO OMAR ESCAMILLA GONZÁLEZ

In 1897 the well-respected civil engineer Antonio M. Anza finished overseeing the construction of the Cabinet and Laboratory for Construction Materials in the Escuela Nacional de Ingenieros (National Engineering School, ENI). It had been a work in progress for fifteen years. In it he showcased samples gathered by other engineers such as Santiago Ramírez. Anza carefully placed the objects in glass cabinets and on shelves specially designed for the collection. The ENI subsequently named Anza director of the laboratory and chair of construction procedures. He retained a prominent leadership role in the laboratory until 1925.

The laboratory was a culmination of past tendencies and also the beginning of something new. The construction materials course that used this working collection was created in large part by mining engineers and mineral assayers who had up to that point produced many of the samples, experiments, and studies about construction materials, yet Anza and the laboratory he oversaw was a catalyst to the rise of Mexican civil engineering. The specimens in this collection provide a wealth of information about construction materials being used in central Mexico during the late

Fig. 2.1. Laboratory and Cabinet of Construction Materials. *Source*: Concepción de Mendizábal, "Memorias de práctica de la alumna Concepción de Mendizábal, hechas en los Laboratorios de Ensaye de Materiales de la Escuela y en el de la Comisión N. de Caminos por acuerdo del Sr. Rector de la Universidad y del Sr. Director de la Facultad" (undergraduate thesis, UNAM, Mexico City, 1930), n.p., Acervo Histórico del Palacio de Minería, Mexico City.

nineteenth century. Contextualized, they also tell us about the attempts of Anza, Ramírez, and many other engineers, geologists, and technical experts to catalog and shape the development of these resources, something they did with mixed success.

The Palacio de Minería, which houses the ENI archives today, served as the home of the ENI from 1882 until it was moved in 1954 to the Ciudad Universitaria, where the Universidad Nacional Autónoma de México (UNAM) is located. The laboratory collection, we believe, had been moved in its entirety to the new campus, where it continued operating until 1962. That year the Society of Engineering Faculty Alumni began a complete restoration of the Palacio de Minería. Subsequently, during the presidency of Luis Echeverria (1970–1976), the laboratory was restored. In 1989 the room with its original furniture designed to store construction materials was given an additional new use; it became the library for the Historical Collections of Palacio de Minería (Acervo Histórico del Palacio de Minería). Archives and a restoration atelier were soon added (see figure 2.1). The library was named the Ing. Antonio M. Anza Library in 2007 in honor of its designer, builder, and the father of construction studies in Mexican civil engineering.[1]

Not all of the materials, however, found their way back into the restoration. Crates of materials remained in a room under the main staircase of Palacio de Minería. In 2000 the director of the Palacio de Minería asked civil engineer Iván Alvarado—UNAM's chair of movable cultural heritage—about what to do with the pieces. He recommended they be kept. He took two pieces to determine their origin, one of which happened to be from Santiago Ramírez, who had provided some of the laboratory's original collection materials, a finding we detail in greater depth later in this chapter. Fifteen years later Alvarado returned to the Palacio de Minería as its chief administrator and asked about what had happened to the artifacts. Having been hired in the interim as archivists, we had, to Alvarado's surprise, no knowledge of them at all. Together we started a fuller investigation of the materials.

After this rediscovery, we took part in an ongoing project that has led to the recovery of approximately one thousand pieces. Searches in our historical archive are still underway to properly account for every piece. The rescued samples will be guarded in the Acervo Histórico for their conservation and further study. We believe they will lend themselves to more research about this and similar collections and their importance for the development of Mexican civil engineering. They will also be of service to architects, engineers, and building restorationists who will be able to compare materials in historical buildings to this and other collections such as the one in the Geological Museum in Mexico City.

In addition to highlighting this important collection, we use this chapter to discuss its origins while placing it within the context of late nineteenth-century nation-state consolidation and globalization. These types of collections started in Europe as materials museums (*matériauthèque*) used for instruction in architecture, for teaching civil engineers, and for creating catalogs for the promotion and sale of materials in national and world fairs. Scientific or "technocratic" diplomats sent word of these museums and fairs back to Mexico where they convinced Mexican academics, government officials, and businesses to participate in their displays. Indeed, Anza himself was the designer of the Mexican pavilion for the Paris World Exhibitions of 1889 and 1900. In this sense, this impetus to catalog and experiment with resources applicable to construction was a means to showcase Mexico's resource wealth and its government's ability to order it, build with it, and sell it. It was a display of Mexican rationality, modernity, and willingness to participate in global capitalism. This endeavor also shows how several Porfirian government officials and members of the scientific community (sometimes the same people) worked together in an attempt to train their own geologists and civil engineers and to create more control over their own resources in order to counter the influence of foreign experts and developers. They wanted to participate in global exchanges, but they wanted to do so in terms more favorable to Mexicans. During the Porfirian era the ENI never trained enough of these experts—nor, for that matter, did the government and private sector leaders ever provide enough support—to successfully catalog all of Mexico's materials or to fully replace its dependence on foreigners. But this process, as exhibited by Mexico's first construction materials laboratory collection and the construction courses that used it, did start it, birthing programs in civil engineering and resource extraction that would become stronger over the twentieth and twenty-first centuries.

ORIGINS OF THE ENI

President Manuel González (1880–1884) gave the ENI its current name in 1882, but it has a much older genealogy. It stemmed in good part from the Real Seminario de Minería or School of Mines, which was founded in 1792 and became the first engineering university of the New World. The school coincided with attempts to shore up the Spanish Empire and then, after Mexican independence, with nation-state construction. The founding of educational and scientific institutions coincided with the establishment of new educational policies and the exploration of Mexico's national geography and resources. Political and economic leaders envisioned bringing social and material progress to the country through these new modern endeavors. In this context, President Antonio López de Santa Anna (1833–

1835, 1837–1839, 1841–1844, 1847, 1853–1855) created the Ministry of Development, Colonization, Industry and Trade (Ministerio de Fomento, Colonización, Industria y Comercio), an institution empowered to affect changes in industry, statistics, roads, exploration and promotion of natural resources, border delimitation, and national and universal fairs, among others facets of development.[2]

The government's scientific explorations were mostly done by the Ministry of Development, and they decided to obtain special commissioners who started looking for iron and coal in 1865. The first engineers the ministry hired for this purpose were Santiago Ramírez, Miguel Bustamante, and Diódoro Leguízamo. They provided a steady stream of reports to the ministry and also to periodicals like *El Minero Mexicano*, Mexico's prominent mining journal, and the *Anales* and *Memorias de la Secretaría de Fomento*.[3] The ministry shared some of its knowledge of the country's mineral resources but also kept some of it classified. Although big efforts were made to cover all territory, the Ministry of Development did not have enough geologists or the budget to map all of the large and mostly arid northern states and territories: Sonora, Coahuila, Chihuahua, Durango, Sinaloa, Nuevo León, Tamaulipas, and Baja California. American entrepreneurs and specialists in part filled that gap, often to their own benefit.

In 1867 President Benito Juárez (1857–1872) transformed the School of Mines into the Escuela Especial de Ingenieros (Professional School of Engineering) after his forces defeated the French Intervention. The opening of the Professional School gave birth to a civil engineering program that built on, while also putting an end to, the combined architecture and civil engineering studies established a decade before at the San Carlos Academy of Noble Arts, the institute that Marcela Saldaña discusses in more detail in chapter one.[4] At San Carlos the Italian émigré architect Javier Cavallari had been teaching civil engineering, but Juarez enacted new education laws that separated the two fields of study, leaving architecture at the academy and moving civil engineering to the Professional School.[5]

This split took longer in practice than on paper. In 1870 Francisco Jiménez was the first person to graduate with a civil engineering degree from the Professional School, but experts coming out of the Academy of Arts were still superior in number and influence; exhibiting this influence was the creation of the Civil Engineers and Architects Association of Mexico (Asociación de Ingenieros Civiles y Arquitectos de México, AIA) that members of the academy had established in 1868. Both schools were important to the early development of civil engineering, but there were also tensions between them, which in some ways hampered a more focused approach.

When President González changed the name of the institution again, this time to ENI, mining engineers and mineral assayers carried out construction materials experimentation because they were the only ones with sufficient knowledge and laboratories to carry out such experiments. This shifted, however, during the government of Porfirio Díaz in the 1890s as demand increased for experts specifically focused on construction with new techniques and materials. Combined with studies in structural analysis and graphic statics, these trends became the foundation of modern civil engineering in Mexico.[6] Engineers working in these areas provided evidence and materials for people working in theoretical physics and mathematics, giving birth to new scientific studies and careers in those fields at the National University for decades thereafter. As an organization of applied science, the budding civil engineering program was further developed by its members' work in Mexico City, where young engineers and engineers-in-training aided foreign and domestic contractors in the construction of massive structures in a city erected precariously and ambitiously over an ancient lake. These studies and practices in building foundations fueled strong programs in soil mechanics and, later, geotechnics.

In general the Porfirian government brought significant changes. They were not immediate, but they did occur relatively quickly. Díaz successfully quelled political unrest and enticed significant US and European investment in railroad lines and port expansion and renovation. European and American products and ideas became more easily and widely distributed. The economy grew significantly, and his government considered engineers a requirement for further modernization. They helped attract foreign investors but would also be essential to local investments and the weaning of Mexican dependency from foreign experts. This drive also led to increases in the importation of construction materials, though this originally occurred mostly in northern and southeastern states, the former because of mining and the latter because they produced vegetal fibers, particularly henequen. The flood of new products and the increasing popularity of world fairs quickly made the study of local construction materials a priority. If Mexico was to rise as an economic and political power, and if it wanted to present itself to the world as a modern nation, its people would have to better catalog their resources, standardize production and quality, and make products more available for local and international commerce.

Because the college was charged with forming new engineers who would bring modernity to the country through railroad lines, ports, potable water, new buildings, and drainage systems, the Porfirian government strongly supported the ENI. Although many colleges were originally under the direction of the Ministry of Justice and Public Education, the ENI was placed under the purview of the Ministry of Development (Secretaría de Fomento) in 1882 because top government officials increasingly saw engineers as valuable tools for exploring mineral resources and creating strategic infrastructure. The Ministry of Development even moved its operations to the Palacio de Minería in 1884, in part because of an attempt to direct the formation of engineers who would be employed by the state. While that ministry was headed by General Carlos Pacheco (1881–1891), exploration commissions scoured Mexico to obtain better knowledge of the territory. The ministry hired several engineers from ENI who combined their exploration activities with academic studies.

According to historian of technology Peter Lundgreen, three ingredients are necessary to develop successfully any branch of engineering: a competent professional school where research can take place, state institutions to use and incorporate engineers, and private companies.[7] For Mexico the school was the ENI; the state institution was the Ministry of Development and then, after 1891, the Ministry of Public Works. The third were foreign companies that built railroad lines and modernized ports, though domestic concerns became more prominent as Mexican engineers increasingly started enterprises on their own at the beginning of the twentieth century.

The Geological Commission, which started in 1888, is a good example because it started with a simple focus: make a precise geological map of the country. Nonetheless, it caused profound changes in scientific practice. As explorers covered new portions of the country, their observations and collection of samples and data gave birth to the study of new branches of geology, such as paleontology. The commission also led to the creation of the National Geological Institute. Mining engineers from ENI were hired as experts who combined their exploration activities with academia, making them geologist-engineers. This duality extended to other specialties of engineering. Geographer-engineers and surveyors were among the founders of the National Astronomical Observatory in 1876 and could be considered astronomer-geographer-engineers.[8] Later on the intense analysis of construction materials and procedures forced civil engineers to study further in higher mathematics in order to calculate complex structures that could function with Mexico City's problematic foundations.[9]

The Ministry was divided into sections in which special commissions were formed to accomplish specific explorative, technical, and scientific tasks. At the end of their work they wrote detailed reports and memoirs, some of which were published in Secretary of Fomento periodicals: *Memorias, Anales,* and *Boletín.*

NEW TEXTS AND TECHNOCRATS

That political leaders believed that better knowledge about construction and construction materials was necessary is clear in the renewed interest in the exploratory commissions and the construction materials laboratory; it was also made apparent by the spread of similar classes elsewhere, Mexican publications on the subject, and the involvement of experts from higher rungs of government. Even while construction materials programs were still led by mining engineers and mineral assayers, courses on the topic were already extending beyond the ENI to the National Agriculture and Veterinary School. This institution was founded in 1853 together with the Practical School of Mines of Fresnillo, Zacatecas. Both schools proposed educational systems based on extensive fieldwork in addition to class instruction. Both agriculture and mining required some expertise in construction.

This emphasis on construction encouraged the publication of several important and influential works in the 1880s and 1890s. Engineer Mariano Bárcena published his *Treatise of Geology* in 1885. He described properties, uses, and locations of the main minerals in the country using theory and suggesting practice to students: "The knowledge of rocks is as useful to a farmer as it is to a civil engineer in order to choose the best construction materials; and also to the mining engineer since it is possible to know the anatomy of the Earth from a geological point of view. This makes it possible to determine the laws governing the location of rocks and mineral deposits."[10] Bárcena was among other things a proponent of lithological geology or the study of the elements constituting the Earth's crust, including those used for construction materials. Bárcena considered granite, syenite, greenstone, amphibolite, and *tezontle* some of the most important stones and spent considerable time describing their physical properties, geographic locations, and uses.[11] As the first textbook on the subject written by a Mexican geologist, the book provided a state-of-the-art survey of the geologic field.[12] The government ordered copies of the book to be used in all college-level geological programs to standardize and, importantly, Mexicanize the discipline.

The first Mexican construction books giving a systematic study of materials and strength coefficients were published in the 1890s. The Ministry of Development printed two texts in 1895. The first was Antonio Torres Torija's *Introduction to the Study of Practical Construction*, written for an applied mechanics course at the National Fine Arts School. It included a table of load compression resistance by square centimeter for quarry, *chiluca*, *recinto*, tezontle, tepetate, iron and brick masonries, *oyamel* wood, and iron.[13] Some of its data were taken from European books by Celestino del Piélago, Julio Jariez, Nicolás Valdés, and Napoléon de Vos. The second was engineer Pablo Argumosa's *Technical Memorandum*, which contained a table of resistances of T-shaped iron beams for permanent uniform loads.[14] It should be noted that although both texts were based on foreign books, they included data from Mexican materials collected by the authors, showing that from the beginning the idea was to shape foreign concepts to fit Mexican circumstances.

Knowledge about construction materials was of strategic importance for the Porfirian state and Mexican exporters and industrialists. Díaz desired a modern country that resonated out from its capital. His government started drainage and drinking water projects in the 1890s. Engineers discovered—though still did not completely understand, as Vitz shows in his essay—how dangerous the unstoppable sinking of Mexico City was and that foundations needed newer procedures and stronger materials such as concrete in order to make a new city with bigger public buildings and monuments. Real accomplishments did happen. Anza saved the Lecumberri prison in 1896 when he used his inverted vaults foundation technique to stop the structure from the destruction caused by differential subsidence. The National Theater (Palacio de Bellas Artes) and the Independence Monument (Ángel de la Independencia) were being devoured by the soil until engineers injected concrete piles to stop it. The specific conditions of the city and the country needed local expertise.[15]

It is also important to note that some of these experts demonstrate an increase in technocratic governance. Bárcena in particular is representative of this growth. He was a pioneer of applied geological sciences in Mexico and was also a botanist. The year the government published his geological textbook, he represented Mexico at the World Fair in New Orleans. He had also become a part of the political elite: toward the end of the 1880s he became a congressman and the interim governor of the state of Jalisco, where he was born. Increasingly certain "scientific" elements of Porfirian circles, and even Díaz, saw technical experts as not only important in the classroom, but also in modern governance.

It is clear that the development of civil engineering coincided with massive changes in Mexico. As engineers gathered and became more familiar with local resources, new building materials entered the country: foreigners brought in steel and concrete, forcing engineering instructors to adapt. Students needed new theoretical tools to start working with innovative technologies, including graphic statics. Neither architects nor practical constructors were particularly skilled in graphic statics, but civil engineers were, making them stand out as the people who could best combine technical knowledge and applied science. As in other modern industrial countries, civil engineers could design structures based on complex physical mathematical calculations and then test the materials in laboratories. Modern Mexico, political and engineering academics believed, would be built by these new professionals.

THE CREATION OF THE CONSTRUCTION MATERIALS COLLECTION

The increased publications and political attention on civil engineering were important to its development, but for civil engineering programs to thrive, students and professors required new collections and laboratories. Intertwined with those creations, they needed hands-on experience in the field. Crews led by guilds had done much of urban Mexico's construction work up to this point, and architects and geological and mining engineers had contributed important studies. But even though architects had been certified by the Academy of Arts since 1786, construction materials and procedures had changed little since the colonial period. As Mexican governments began to push more ardently for modernization in the 1860s, officials argued that these conditions could not continue. Under the direction of the academy's program chair and mineral assayer, Agustín Zamora, a new civil engineering program created Mexico's first construction materials course (Conocimiento de Materiales de Construcción) on May 15, 1869.[16] But despite initial enthusiasm it soon had to close its doors because there were no actual laboratories or collections for teaching the students. Following the wars over the French intervention, the capital city had not suffered from a lack of ideas, but it did lack funds, equipment, sufficient fieldwork, and stability. This period was especially turbulent, but similar issues continued into the Porfirian era. A lack of stable funds and support hampered the development of construction studies and, subsequently, civil engineering. Nonetheless, significant progress was made.

Although the Ministry of Development gave support to renewed progress on the construction materials program at the ENI, during the early 1880s most experiments on construction materials were still being done by mineral assayers and smelters, whose professional work was closer to mining and metallurgy than civil engineering. Mining engineer and director Antonio del Castillo was able to acquire some new equipment, but more than anything else, the ENI leadership focused on chemistry labs, not civil engineering.[17] The lack of attention and laboratories left civil engineers out of mineral assaying until the 1890s when economic and educational policy shifts led to the creation of the Ministry of Communications and Public Works, which became an institution better suited for civil engineering skills.

It was under the direction of engineer, diplomat, and ENI professor of materials Gilberto Crespo y Martínez, beginning on February 14, 1882, that the school formally reinstituted the course on construction materials. A technocrat of diverse talents, Crespo y Martínez was a professor of mineralogy, geology, and paleontology. He was immensely interested in construction materials. The way he saw it, "no engineer can become a good builder if he ignores the materials used to build."[18] By this point, the course was divided into seven parts. The first two were focused on principles; the rest were devoted to the most-used materials—mainly iron "because of its importance in construction," and aluminum—and then studies in practical construction.[19]

The drive to move forward with the program came in part from technocratic diplomats in Europe. Engineer and scientist Francisco Díaz Covarrubias, who helped obtain equipment from France in 1886, had earlier become aware that construction materials displays and research cabinets were becoming commonplace in Europe. Geologist Felix Karrer, for example, had been gathering samples of materials used in all important buildings in Austria-Hungary and in other prominent buildings across the continent. Each specimen consisted of a cube that he provided to the mineralogical section of the Natural History Museum in Vienna.[20] These collections were proving important to studies in European construction, even though many scholars in the region still saw the samples as materials for the earth sciences, not specifically for civil engineering programs.

The French Ministry of Public Works had started a collection of stone construction materials for the 1878 Paris World Exhibition.[21] The goal was to present a six-part commercial catalog for visitors. First, there were 26 big carving stones from the main quarries in the country. The second section contained 677 samples, also carving stones, of 24 × 24 × 16 cm blocks. The third component consisted of 123 12 × 12 × 8 cm marble samples. The fourth part consisted of 99 plates, also of marble. The fifth section held 141 small containers of limes, types of cement, and plasters. The sixth and last part presented samples from thirteen pavement, twenty-seven brick, and seven clay producers. It's very likely that engineer and diplomat Gilberto Crespo visited the exhibition, saw the collection, and reported it to the Ministry of Development. As previously stated, the ENI shortly after declared that Mexico needed its own Mexican construction materials collection to support the teaching of civil engineering programs.[22] In 1886, following initial collecting expeditions and a brief defunding of the laboratory, a 2,267.50 pesos budget intended for the construction materials course excursions was instead spent in Paris. The Mexican general consul in that city, Covarrubias, used the money to buy machines for a new laboratory.[23] It had become clear that for a course in construction materials—and the civil engineering program in general—to be truly successful, it would first need more space for experimentation and a collection of cutting-edge, though increasingly prevalent, construction materials such as clay bricks, iron, steel, and cement.

Fieldwork on the ground pushed government officials to buy this equipment. The leadership of the Ministry of Development and ENI directors—first Antonio del Castillo and then Crespo y Martínez—commenced the construction materials collection in 1882. The Ministry of Development decreed its support for gathering these materials, natural or artificial, as valuable to the strategic development of the nation. Del Castillo wrote a letter on January 28 to Secretary Carlos Pacheco asking him to order all road crew leaders, telegraphic line directors, and railroad inspectors to start forming "collections of construction materials used in the states, districts, and territories where they were as soon as possible to be sent to this school through the ministry."[24] Ministry of Development Employees (state engineers) also directly collected specific and strategic information because the work required to build the collection superseded the number of available inspectors. Mining engineers, for example, were required to do research on local construction materials and collect them while they were prospecting. A similar request was made of civil engineers working on railroad lines or port construction. These requests brought in the initial materials for the construction materials laboratory and course.

Collecting protocols stated that wooden samples should be of 0.15 m length, 0.08 m width, and 0.03 m depth to make them manageable in the cabinet. Del Castillo also wanted them labeled with the wood's common name, collection locality, and its prescribed uses. For rocks he wanted the blocks to be 10 cm cubes with a similar label. Special attention was to be given to localities with foundries, where construction materials such as iron, lead, and copper were collected. Finally, railroad enterprises had to send samples of all types of their rails in use.[25]

Del Castillo's request was based on a nearly total lack of information about the properties and uses of Mexican materials at the ENI. He emphasized how important it was to do resistance experiments to know which of them could be exploited with commercial success. He even demanded relocation of a wood collection owned by the government of the state of Veracruz to the ENI (showing that some states had already conducted some of this type of work), since it was ready for the students.

The Ministry of Development issued two orders on February 14 and August 25, 1882, to formally start the collection for the construction materials course at the ENI. Measurements and data varied in the original request for the involved materials. The ministry wanted cubic samples of 0.3 m and a label not only with the common name of the rock, plaster, or lime; it also gave special instructions for bricks and tiles to include the location, their use (carved or block), measurements, prices, exploitation techniques, the ovens used for brick and tile firing, and a sample of the sands used in the process.[26]

The first three shipments of materials in 1882 carried out by special commissioners of the Ministry of Development exemplify how the system worked. The first delivery was sent by surveyor José Antonio de la Peña y Ruiz, the director of a route of study from Jalapa to San Marcos, near Coatepec, in the state of Veracruz. His crew sent two boxes filled with wooden samples. Professor del Castillo approved of the samples, which he thought would be "useful to the advancement of sciences taught on the institution."[27] Unfortunately, any description of the materials has been forever lost or has yet to be uncovered. The second shipment was made by José María Velázquez, a surveyor and civil engineer who had become inspector of the Mexican Central Railway between the cities of Querétaro (state of Querétaro) and León (state of Guanajuato). He sent a collection of rock cubes from the state of Guanajuato using the dimensions suggested by del Castillo which arrived at the ENI on June 6, 1882.[28] The third and last shipment of that year was a collection of construction materials used in Ixtapan de la Sal, in the state of Mexico about 70 miles southwest of Mexico City, sent to the secretary of development by mining engineer and special commissioner Santiago Ramírez on December 11. This collection has proven the most important for archivists and historians today, and likely for the Ministry of Development then, because Ramírez provided rich descriptive details, which he included in an account titled "Notice of the Rocks and Materials of Construction Collected in Ixtapan de la Sal and Shipped to the Secretary of Development." Ramírez states, "I should start by establishing that giving the samples the specified dimensions [on the secretary's order] led me to find some troubles inherent in the rock's texture and extraction, and these [factors] make their transportation very difficult, particularly on those winding paths where only beasts can carry them."[29] Among the shipped samples was a rock of little density which weighed eighty-one kilograms and was nearly three cubic meters in size; that is, it was very nearly the load of a donkey.[30] Few of these materials were known outside the region where they were produced; thus collections such as this were immensely useful for Mexico City engineers and technocrats who used them to train in construction materials but also to catalog, make legible, and hence attempt to wield some control over regional resources.

Shipment transportation such as Ramírez's large boulder was expensive and slow, so he decided to reduce the samples to 10 cm cubes. He took the license to write reports instead of labels because the required information was more than could feasibly fit on a label. One existing report recently uncovered provides an account of seventeen collected samples. He gave their common names, a description of their locations, their uses, whether they were carved or natural, their sizes, their prices, and how they were commonly exploited. The report is so detailed that it allowed us to easily locate every point on a present map and thus confirm that the samples were taken out of quarries found on shale formations near Ixtapan.

Ramírez describes "over the cinder dominating the formation of Ixtapan, that can be seen on peaks standing out of the soil of the town and its outskirts, there is a big extension of clayey slate impregnated by quartz."[31] His first two samples were *canteras*, which he described as "a feldspar porphyry, barely altered by the decomposition of feldspar." The only difference between them was their color; both were taken from a deposit "10 kilometers away from Ixtapan, 4° to the Northwest" between the towns of Plan de San Miguel, Yerbas Buenas, and Ahuacatitlán in the state of Mexico.[32]

Ramírez declared that these quarries were used to build the churches of Ixtapan de la Sal and nearby towns, particularly those whose residents needed solid but also carvable stones to make ornaments. His third sample came from Ahuacatitlán and is similar to the previous ones. The fourth and last of the quarry samples was called "firestone" (*piedra de fuego*) because it was used to build boilers and furnaces.[33] He also explained that the collected samples were exploited with rudimentary and imperfect technics with a crowbar and that small, imperfect fragments were rejected in favor of bigger, regular ones that were suitable for construction. Their common size was 40 cm long, 25 cm wide, and 20 cm thick.[34] As Rocio Gomez demonstrates in her chapter, many locals relied on customs instead of new state policies and tools that they either resisted or had not encountered.

After the fourth sample Ramírez focused the rest of his attention on stone deposits used in the manufacture of lime on a rock formation called Cerro Alto. He collected samples of a gray bluish limestone that was easily found at the place. He collected "burnt" materials, that is, the final product.[35] The mixture used for this construction material consisted of lime and sand. He collected more of these finished materials from nearby locations, including El Mal Paso and Cerro Alto in Ixtapan, El Potrero de Jesús in the nearby district of Tonatico, and two samples from Paso de Guadalupe.[36]

The last five of his seventeen samples were related to the manufacture of bricks, tiles, and pottery. Ramírez collected clay pieces used for this purpose in the towns of Ixtapan and Tecomatepec. All of these products were used extensively. Decades later in 1970 similar clay vessels were manufactured in the area, although with poor organization and in small numbers.[37] Similar pottery has become popular again because tourists view it as representing something authentically Mexican.

As for manufacturing processes for bricks and lime, Ramírez stated that furnaces were so similar to those used in Mexico City that it was useless to describe them in detail. Torres Torija described the facilities, tools, and steps needed to obtain the finished products in his late nineteenth-century textbook, but he failed to include images, thus making it difficult to imagine exactly what Ramírez had seen in Ixtapan.[38]

Ramírez concluded his report by stating that prices for a thousand bricks were between 7 and 9 pesos, and for one *carga* (11.5 kg) it was 20 to 22 reales (160 to 176 pesos).[39] According to Torres Torija's and Argumosa's texts, selling conditions and prices for Mexico City were totally different. For one thousand bricks, the first author refers to a price between 10 to 15 pesos, while the second gives 10 to 12 pesos, both estimates dating from

Fig. 2.2a. Ixtapan de la Sal brick collected by Santiago Ramírez in 1882. The brick has two glued paper labels. The one on the top states: "Construction materials no.12 Brick from Ixtapan de la Sal District of Tenancingo S. R." All labels were handwritten by Ramírez, as shown by comparing this writing with his signature. *Source*: Acervo Histórico del Palacio de Minería, Mexico City.

1895.[40] For lime, the differences were even greater. Torres Torija wrote that one *carretada* of 120 arrobas (1380 kg) was sold at different prices depending on the season, because rain made it difficult to manufacture consistently, but that it would normally cost between 25 and 45 pesos, even though it could reach between 80 and 100 pesos when availability was limited.[41] The greater demand and commerce usually created higher prices in the capital. The aforementioned carga of bricks could cost the equivalent of 160 pesos, but it could rise as high as 288 pesos, far higher than Ixtapan.

Argumosa provided further notices specifying three types of lime: *molonque* (powder), white, and fine, also sold by the carretada. The price range for one carga was 96 to 160 pesos, 128 to 192 pesos, and 128 to 224 pesos respectively.[42] Comparison of the data given by these three authors shows large discrepancies in quality and quantity of the produced construction materials between Mexico City and other regions, even in relatively close rural areas in the state of Mexico. Historians should focus attention on these differences if a better understanding of construction technologies and economics is to be more clearly understood. Other factors that need to be taken into account are the construction of railways and the mod-

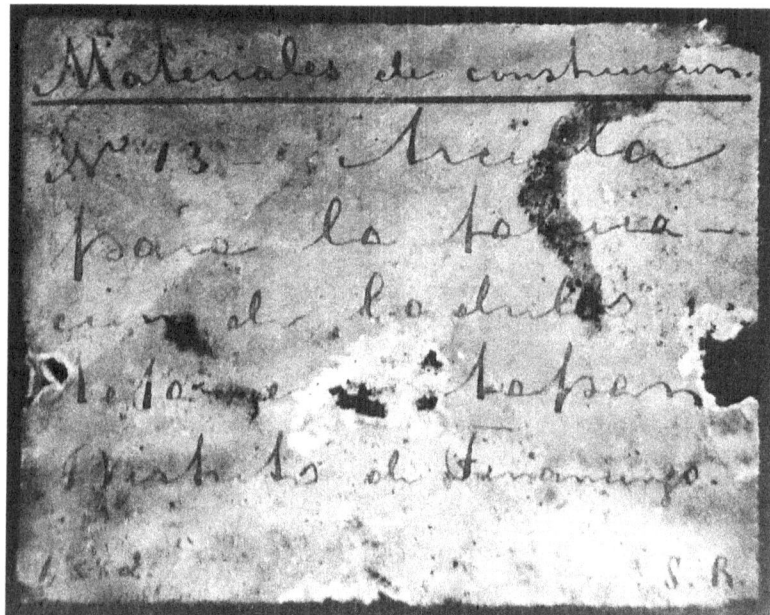

Fig. 2.2b. Clay used for manufacturing bricks in Ixtapan de la Sal. Sample collected by Santiago Ramírez in 1882. The photograph was made with UV light because inks made of iron and gallium are fluorescent under ultraviolet radiation, making faded writing easier to read. *Source*: Acervo Histórico del Palacio de Minería, Mexico City.

ernization of ports that made international commerce easier and provided increased materials from New York, Chicago, and New Orleans to places including Tamaulipas, Veracruz, and Yucatán. These changes often introduced foreign materials to regions outside of Mexico's central valley where foreigners first reached ports of entry.[43]

One of the more important takeaways from the Ramírez materials is that they show Mexico was not unified in its construction standards.[44] The most commonly used brick in Ixtapan was based on the Spanish *vara* (84 cm) and was 12 × 12 × 8.4 cm (1/7 × 1/7 × 1/10 varas) in dimension, though the cuts were not always precise (see figures 2a and 2b).[45] However, the common brick in Mexico City at the time—at least as described in Mixcoac, Mexico City, by Torres Torija—was 28 × 14 × 3.5 cm (⅓ × 1/6 × 1/24 varas).[46] Material sizes were not standardized even though the government had made several attempts to introduce the metric system.

In 1970, eighty-eight years later, engineer Adrián Aguilar Contreras visited the same region. He wrote some reports as commissioner of the Directive Committee for the Research of Mineral Resources of Mexico (Comité Directivo para la Investigación de los Recursos Minerales de México), a state institution created after the Mexican Geological Institute was decentralized and given to the National University. Later it was reformed as the Mexican Geological Survey (Servicio Geológico Mexicano) under the direction of the Ministry of Energy, where it remains today. Aguilar was sent to study pottery and brick manufactures in Ixtapan de la Sal.[47] This means, that two geologists who made the same excursion almost a century apart were both paid by the government to explore natural resources for manufactures related to construction and ornaments in this region well known for its production of clays and metals. This report mentioned little of the use of bricks in the region. Locals now mostly used clay in pots for domestic use and sale to tourists. The production of bricks had noticeably lessened, and the size had changed to the more standardized proportion used in the capital and across much of the country.

In some ways the two reports show the inefficiency of the Mexican state in organizing, maintaining, and using this type of scientific and economic evidence gathering. The author of the second report never even knew about the first one's existence. A lack of consistent funding and leadership—after all, a revolution separated the two reports—made the exploitation of strategic information difficult. Neither the state nor business leaders capitalized on building a brick industry out of the Ixtapan resources. That is not to say that different government officials and industrialists were not interested in supporting these types of industries, more that reports of these producers and goods did not always make it into the hands of top leaders or got lost in the changing bureaucracies and shifting governments, sometimes only to be found by archivists more than a century later. It is even more likely that poverty, migration, foreign products, and a lack of national investors had dulled interest in these industries and allowed them to wither.[48] Nonetheless, it is clear that the works of technical experts and Mexico's increasing involvement in global capitalism had a rationalizing effect. After all, the bricks in Ixtapan in 1970 were similar to bricks across much of Mexico and, for that matter, much of the world.

The study of materials that were subsequently placed in the construction materials laboratory would go on to substantially alter Mexican development. As modern civil engineering was born in tandem with a better understanding of Mexican geology, the standardization of building materials, and, importantly, the introduction of concrete foundations, Mexican engineers more clearly brought to light the problem of soil compressibility caused by erecting Mexico City over a lake. Incorporating local circumstances and materials in addition to foreign technics and materials, engineers tried to solve this problem, leading to important discoveries in soil mechanics.

The resulting solutions caused their own problems, but they also allowed for the construction of skyscrapers such as El Moro (national lottery), La Nacional (insurance company), and Torre Latinoamericana (insurance company) in the 1940s. During the industrial and economic development in Mexico known as Milagro Mexico o Desarrollo Estabilizador (Mexican Miracle or Stabilizing Development) between 1940 and 1970, engineer Bernardo Quintana Arrioja (1919–1984) founded a company called Associate Civil Engineers (Ingenieros Civiles Asociados, ICA) in 1947. It became an official state contractor for fifty years and donated its research laboratories to UNAM. This successful project by entrepreneurs, the state, and academia helped the development of a class of Mexican geotechnics experts who have become renowned around the world.[49] These endeavors are historically connected to the rise of construction studies and civil engineering in Mexico. It is not hyperbole to state that the origins of these studies and accomplishments can be traced back to the foundation of the construction materials laboratory.

CHAPTER 3

THE PREOCCUPATION WITH SAFETY

Mining Engineers, Education, and Practice in Modern Mexico

ROCIO GOMEZ

In the extraction industry miners' bodies and their tools served as the machines that hollowed the earth in search of precious metals as Mexico moved into the twentieth century. Long a profession but only recently subject to professionalization, mining engineers in Mexico sought to remove the ore from mines efficiently and then to convert it to precious metals through a complex chemical process. By the late nineteenth century miners began demanding basic occupational safety measures and saw them codified into formal mining laws. Mining engineers received the added task of protecting the lives of miners as outlined in those laws. While the laws were a step forward for occupational safety, how mining engineers implemented such measures underscored the obligations mine owners and mining companies had to their workers. With Mexican lands and mineral rights subject to foreign exploitation in the late nineteenth century, miners and their bodies became the subject of discussion as mining companies weighed new standards of safety in the mines as well as how to best respond to occupational hazards. Coupled with the changing standards of professionalization, Mexican engineers attempted to create an illusion of safety for miners in an era of rapid growth in the mining sector.

Mexican miners have always faced perilous journeys underground every time they clocked in for daily shifts. However, in the late nineteenth century they faced the added peril of the introduction of new technologies from foreign investors. Pneumatic drills, heavy explosives, and a greater excavation push into old mines all served to rattle the nerves of the most seasoned miners. How could mining be made safer? By what standards should the industry be made safe? What was the end goal of creating occupational safety guidelines? To achieve some success in curtailing accidents, mining engineers needed to receive training that went beyond the theoretical. Typically trained in classical mathematics and physics, students of mining engineering regularly lacked sufficient hands-on expertise in implementing safety designs in the mines. Further, they often found miners with years of experience and informal knowledge who were far more skilled in preventative measures for occupational safety.

This chapter argues that mining engineers attempted to limit occupational accidents in the mining sector during the Porfiriato. It demonstrates that awareness of safety measures began as informal knowledge and was largely experiential before becoming standardized to a degree in the early twentieth century. Mining engineers obtained a growing number of opportunities for professionalization through education and apprenticeships, which incorporated safety training. A combination of worker interest, mining codes, and federal legislation attempted to hold engineers accountable for their efforts in protecting workers. The miners and mining engineers who injured themselves or died on the job, presented here as case studies, represented the monumental efforts needed to protect miners from accidents. In monitoring an inherently dangerous profession, mining engineers had the guidance of mining codes and a solid classical education but lacked actual experience in protecting workers, which was a calamitous combination for miners who counted on these engineers for basic protections. Worse still, miners saw the resulting medico-legal discourse around occupational safety default to mining companies' favor. This chapter first explores the education and professionalization of mining engineers. It then uses individual case studies to show some of the difficulties they faced during their attempts to implement safety measures.

While historians have centered on Mexico City as the core of engineering education, miners dwelled in smaller cities such as Zacatecas, Guanajuato, and Pachuca. This chapter focuses significantly on Zacatecas, showing how engineering knowledge was applied in regions outside the capital. These mining cities exemplify the extent and the limits of reforms as they grew during the Porfirian era into revolutionary endeavors. They also show the national capital's emphasis on consolidating education, money, and power.

These trends can be seen in the role engineering played in formulating occupational health and safety. Engineers, and particularly mining engineers, attempted to balance extraction with the maintenance of miner safety. By checking the mine's internal structures, supervising the carrying of ore, and facilitating a safe environment, they devised the most productive system to optimize extraction. But the profession as a whole did not seem overly concerned with the consistency and effectiveness of occupational health and safety measures, until these measures were required by law. Even then, engineers in Mexico implemented safety measures inconsistently, with inspections occurring only every two years by the late nineteenth century. In short, miners and other mine workers often served as the first resource for identifying occupational hazards. Only through pressuring supervising mining engineers did they receive any protection.

A QUICK HISTORY OF OCCUPATIONAL HEALTH AND SAFETY

Historians and mining scholars have centered their concerns for the occupational health of miners in classic European works going back centuries. Georgius Agricola (1494–1555) noted how miners developed a persistent cough or suffered terrible accidents while on the job. In Agricola's assessment, underground fires, dusty environs, and shoddy construction all contributed to the short life span of miners.[1] Occupational disease and workplace safety remained a subject of interest for medical students through the early modern period. In the late seventeenth and early eighteenth centuries, Bernardino Ramazzini (1633–1714) methodically examined proto-industrial sectors to underscore how the occupational disease affected individual trades, providing a model for how engineers could observe the physical act of work. Notably, he began his examination with miners, citing their importance as well as the saliency of their ailments. Miners suffered from lung disease, and they contended with the daily threat of imminent collapse in the mine. Ramazzini wrote that the "metal-diggers" suffered from the "noxious quality of the matter" handled in their daily actions, which in turn affected the "natural structure of the Vital Machine" or, rather, their human body.[2] He enumerated their diseases, citing "difficulty of breathing, phthisis, apoplexy, palsy, cachexy, swellings of the feet, falling of the teeth, ulcers of the gums, pains, and tremblings in the joints, so that upon the whole their lungs and brain are affected."[3] In short, the list of ailments painted a picture of a "Vital Machine" broken by wear and exposure to a demanding profession. Notably, even in rhetoric Ramazzini imagined workers' bodies in engineering terms: a machine capable of breaking down as well as subject to physical and chemical changes over time.

How did engineering play a role in formulating occupational health and safety in Mexico? Engineers, and particularly mining engineers, attempted to balance the operations of mines and the safety of miners. By checking the mine's internal structures, supervising the extraction of ore, and facilitating a safe environment, they devised the most productive system to optimize extraction. At least that was the goal. In doing so mining engineers transformed the human bodies of miners, making an external structure (the mine) inform the internal structure of the human body.[4] Dust kicked up in the air, slippery slopes, constrictive spaces—they all shaped miners' bodily experiences. The purpose of mining engineers was to assure the continuation of extraction, and hence protecting the health of miners seems like it would have been relevant. Miners, after all, facilitated the removal of valuable ore from underground.

Miners and their bodies illustrated the effects of labor on the human organism, which made the mining sector a catalyst for global conversations regarding occupational safety, especially lung diseases. By the nineteenth century mining safety had become a topic of conversation in mining communities around the world. Historian David Turner has examined how miners' bodies became a topic of discussion in newspapers and magazines in Britain.[5] Journalists wrote during the Industrial Revolution of coal miners covered in black dust with hunched backs to illustrate the transforming and deforming capability of this type of labor. While the external effects of this labor were visible, other ailments worked their way into miners' internal organs, namely the lungs. While silicosis and other lung diseases did not properly receive their modern name until the 1920s, miners exhibited all of the classic symptoms of occupational lung diseases: coughing, listlessness, limited mobility, and breathlessness. From silicosis to asbestosis to coal miners' pneumoconiosis (black lung), respiratory diseases associated with mining engendered global conversations in the early twentieth century. In the United States coal miners spurred legislative action through their unions in order to understand the effects of their symptoms on their health, their community, and their future.[6] Meanwhile, in South Africa gold miners received the first compensation for silicosis tuberculosis in the 1920s, albeit through a very restrictive and racist system.[7] By the 1930s and 1940s the idea of compensation for occupational health became commonplace, even in Latin America. Historian Ángela Vergara, for example, has examined the role of Chilean labor unions in advocating for worker protections and copper miners afflicted with silicosis.[8]

In sum, miners around the world interacted with separate legislative and medical systems to advocate for safety measures. This raises the question: What role did mining engineers have in this new global concern regarding miners' bodies? At least for Mexico, the answer lay in engineering education.

EDUCATING ENGINEERS AND CENTERING THE BODY

Before 1793 miners of Indigenous, mixed-race, or African origins often held much of the mining knowledge in cities such as Zacatecas and Guanajuato. Both freed and enslaved miners informed the trajectory of mine expeditions underground and helped shape mining methods, especially in the amalgamation process. Amalgamation, the process of drawing out silver from the raw ore, involved an intricate balance of supervision, timing, chemistry, and the frequent mixing of the compounding material. Mixed in large round *tortas*, or circular patterns, on an open-air patio, the mixture was attended to by patio workers and *azogueros*. These workers specialized in mixing mercury (Hg or *azogue*) into pulverized silver ore on large patios to "pull" the silver from the mixture. Baking, washing, and polishing followed to complete the process.[9]

Azogueros, miners, and patio workers alike contributed their informal knowledge that fed broader imperial imperatives regarding production. Built on years of experience, they frequently used or defaulted to methods sometimes deemed wasteful by engineers such as Friedrich Sonneschmidt. At the request of Fausto de Elhuyar, director of the Real Seminario de Minería, Sonneschmidt (1763–1824) worked to improve Mexican amalgamation beginning in the 1790s. The mining engineer argued that the waste of the azogue per each amalgamation effort had been a frequent source of discord between azogueros, with some arguing that the loss was inevitable.[10] Yet, as literary scholar Alison Bigelow notes, mercury from China and Europe brought to the Americas allowed trained engineers to implement and prioritize new methods, superseding previous ones involving localized knowledge: "The extent to which officials in New Spain invested in mercury suppliers and distribution networks in Europe and Asia underscores an important part of the history of American silver . . . [N]ew metallurgical methods . . . made it practical to refine silver and position it in global markets, and [built] economic and information networks that were credible enough to sustain global trade."[11] For Crown bureaucrats, the demand for precious metals fueled the drive to incorporate new methods of extraction. Future research, in the vein of Bigelow's work, should prioritize Indigenous and African knowledge regarding mining and amalgamation methods to understand the broader impact of informal scientific knowledge on local and global silver markets.

In the face of an increasingly global trade, Spanish bureaucrats in the late eighteenth century invested heavily in the creation of new institutions focused on formal and technical engineering knowledge. Don Juan Lucas de Lessage, the *regidor* of Mexico City, and Don Joaquín Velázquez de León, a mathematics scholar and royal counsel, expressed their dismay regarding the Mexican mining state of affairs in 1774.[12] "The guild or community of the miners in this New Spain, is a Body (if it merits the name) without a head, lacking organization; but whatever species of man, that they are all of the same profession, exercise, or destiny in the order of human society, it needs a homogenous head, that could direct its members with intimate knowledge of true and proper nature. . . . Being our miners independent and dispersed men, each one preoccupies himself with his own business, and no one solicits or knows of, the general interests of mining."[13] The attempt to consolidate mining knowledge under one institutionalized body hinted at colonial anxieties. Miners were dispersed and not subject to any type of checks. They operated largely independently, according to de Lessage and Velázquez de León. However, their assessment largely ignored miners who were enslaved or in otherwise forced labor. In addition, both bureaucrats argued not so much for an institutional leader, but rather a consolidation of methods of production by trained engineers. Because efficient methods meant better production results for the Crown, they sought to introduce these methods through trained individuals rather than by chance.

Whereas most mining knowledge relied on locals with generations of practice, de Lessange and Velázquez de León found themselves attempting to professionalize the fields through new institutions. In 1792 the Real Seminario de Minería (later known as the Palacio de Minería) was established in Mexico City under the supervision of Fausto de Elhuyar (1755–1833), a Spanish chemist charged with staffing the new institution. This period coincided with a scientific awakening in geology and refining methods in the region, during which the Crown invested in mining research to increase the output and quality of precious metals in the region. Engineers, such as Miguel Bustamente, gave public and invited talks at the Real Seminario de Minería, and new trade publications disseminated novel amalgamation techniques, vein tracing methods, and discussions on potential lodes.[14] The overall goal was to keep these bright minds from fleeing to Spain, Germany, or France for training and instead to "grow" these engineers in Mexico.

Professors and students at the Real Seminario focused on traditional mathematics and sciences. The founders referenced occupational safety only in passing. Viceroy Miguel de la Grúa Talamanca of New Spain received a message from Crown ministers underscoring the importance of the Seminario as they "learned the theoretical, practical, and appropriate knowledge for the most secure and efficient labor in precious metals of gold and silver in those Royal Mines."[15] This subtle nod to safety underscored the safety only of the metals and not so much of the workers. At the same time it did acknowledge the need for practical understandings of mining in the education of future mining engineers, which also became a critical part of engaging more with occupational safety.[16]

During and after the Wars of Independence (1810–1821), faculty at the various schools scattered or returned to Europe to flee anti-Spanish legislation and retribution. Meanwhile the Palacio de Minería continued to offer courses largely composed of the traditional sciences. Faculty did not provide a curriculum for applying technical methods in practice, which proved detrimental for occupational safety. Students calculated the physics related to how much weight a beam could carry but did not know the mechanics of the mine and the circumstances that could affect such calculations (e.g., humidity, heat, geological fortitude of sidewalls, etc.).

Student engineers also fell victim to the fickleness of Mexican politics and regional rivalries, creating an inconsistency in their education. Liberal and conservative presidents in the mid-nineteenth century weathered foreign interventions, secessionist crises, military invasions, and political changeovers. Each conflict unsettled and disrupted meaningful progress in the education reform of engineering schools by sapping funds or conscripting young male students.   Education also became subject to petty rivalries between Mexican states, a hallmark of nineteenth-century Mexico. In 1853 President Antonio López de Santa Anna granted a charter for the Escuela Práctica de Minas y Metalurgía, or Mining and Metallurgical School, in Fresnillo, Zacatecas. The charter created a regional school for the "theoretical and practical training" of mining engineers.[17] Student engineers and miners attempting to certify their training or climb the ladder of professionalization in engineering finally had a regional education center that combined education and practical experience. However, in 1857 politicians in the city of Pachuca, Hidalgo, demanded the relocation of the school to their town. The Ministry of Development had confirmed the transfer of the school to the new city. An opinion piece in the newspaper *El Eco Nacional* stated that the move was not necessary and would only be wasteful because a building in Fresnillo had already been built specifically to house the school of mines, at the cost of 60,000 pesos. In a sly dig, the authors expressed their doubt that Pachuca had "the rich variety of minerals as in Fresnillo and which are so important for the best instruction."[18] Ministers ended the tension with the closure of the school in 1861.[19]

Mining engineers typically earned the title of *ingeniero* through formal education or through experience, a difference that underscored attempts at professionalization during the nineteenth century. Even by the late nineteenth century, many engineers did not have a degree declaring them formally as ingeniero. They typically learned through more informal experiences in mines in a particular area. These engineers-in-name-only then often worked in peripheral mining cities in the northern and central states, far removed from the elite circle of the Palacio de Minería. While some students came from these hinterlands, engineers produced by the Palacio often returned to work the mines near their homes.

EDUCATION AND PROFESSIONALIZATION IN MODERN MEXICAN ENGINEERING

During the Porfirian era engineers increasingly adhered to specific educational requirements that grew to meet the demands of political leaders. In an attempt to support public health, *científicos* in Mexico City inspired a managed approach to sanitary conditions throughout the country. These technocrats attempted to curb seasonal floods and sewage overflows that had plagued the city since the arrival of Europeans. Newly degreed civil engineers and students assumed the tasks of designing fountains to distribute potable water to neighborhoods while also building the Gran Canal del Desagüe, which carried wastewater away from the capital toward the Gulf of Mexico.

A different type of flood came to the mining sector. With the Mining Law of 1884, foreign investors arrived in droves and created a demand for mining engineers. British, French, US, and other investors then purchased mineral rights, while the subsequent law of 1892 granted water as well as mineral rights to foreign owners. Speculators scrambled in the buying frenzy to hire mining supervisors who had studied mining operations, recognized geological formations, and implemented basic safety measures. According to the *Boletín Oficial de la Cámara Minera de México*, only 109 engineers graduated with a specialization in mining from 1859 to 1909, a "sarcasm in a country that declared itself as a mining center."[20] In response to the investment in mining and the lack of mining engineers, the Cámara Minera, the government body supervising the extractive industry, outlined new requirements for engineering schools to best prepare students for the mines.

Mine inspectors, typically associated with municipal mining councils, served as the foundation for occupational safety practices. While their visits served as initial reviews to see if the mine was indeed operational, they also checked to see if basic safety protocols were being observed. Inspectors had checklists on their visits to the mines with the expectation that they observe the conditions of miners, fortifications, and ventilations.[21] In most cases, they provided the only official review of the structures supporting the mine walls and miners themselves. These structures, typically made of wood, were rarely built by engineers, but rather by miners or other workers familiar with the needs of the mine. Inspectors also monitored the apparatuses facilitating extraction and transportation: interior winches, electrical wiring, cables, ladders, and cart tracks. Because of this hefty assignment, the government required structural as well as electrical engineering basics to be common knowledge among inspectors. While structures in the mines emerged from intergenerational discussion among mine workers, electricity offered insight into a new phenomenon in engineering, particularly for inexperienced employees. This new field required special instrumentation and handling of potentially dangerous items. Amateurs without training quickly found themselves outpaced by technology.

During the Porfirian era engineering schools increasingly emphasized professionalization through a classic engineering curriculum *as well as* firsthand experience in mines. Secretaría de Fomento Manuel Fernández Leal chaired a committee that supervised a curriculum overhaul in engineering in 1883 and argued for engineering students to be directed toward a "mechanical rationale" in their academic principles and a curriculum that emphasized classic engineering.[22] Faculty at the Escuela Nacional de Ingeniería agreed to this targeted approach, restricting German literature classes to only the introductory levels, suppressing the studies of literature and history, and adding a course in mineralogy and geology. Furthermore, the Cámara Minera made linear drawing a requirement for the sake of *croquis*, or engineering schematics.[23] Fernández Leal went further and suggested Spanish language and literature be eliminated from the curriculum as well. Sr. Guillermo Puga, a member of the curriculum committee, contested this advice instead of arguing for their inclusion. Committee member Juan N. Anza agreed, positing that a student might take Spanish literature and determine they disliked engineering since the end result of education is for "social and individual perfection."[24] According to historians Gerardo Tanamachi Castro and María de la Paz Ramos Lara, these conversations attempted to streamline the course requirements for the degree of *ingeniero de minas*, but the curriculum was slow to emphasize physics.[25] As they note, the inclusion of physics into the curriculum at the Escuela Nacional de Ingenieros coincided with the country's process of modernization. Moreover, the physics courses changed in accordance with what the country needed. Courses in electricity and magnetism coincided with the introduction of the electrical grid. Machinery courses coincided with a growing mid-twentieth-century push on machinery. Despite the curriculum changes in 1883, the committee underscored students' ability to apply and practice what they learned in the classroom.

Fernández Leal and Puga quarreled over the curriculum balance between theory and practice. The Cámara Minera noted that the courses on chemistry, mathematics, and physics provided a foundation but continued to lament the lack of practical education.[26] Mining engineering students employed considerable theory and calculation: from the stress of joints to the analysis of geological formations underground to knowing the weight limits of different winches. Yet students did not have firsthand experience in applying complicated chemical, physical, and metallurgical techniques in transforming raw ore from the mine to precious metal. In addition, among these course changes prospective engineers took courses in metallurgy, followed by specific metals, such as iron, steel, copper, lead, silver, and gold. Adding to the list of courses, engineering students in the latter years of their curriculum at the National School of Engineers in Mexico City undertook in 1909–1910 courses entitled Theory and Practice of Labor and Mining Administration and Subterranean Topography.[27] Professors and administrators took students, upon completion of these courses, on excursions to make coursework applicable to real-life scenarios. Excursions to the mining regions of Sonora included a carbon-ferrous mine in Coahuila, a mine in Cananea, and a site owned by the Rey de Oro Mining Company near Magdalena, Sonora.[28]

In the wake of the Cananea Mine Strike in 1908, which foreshadowed the coming Mexican Revolution, the government became preoccupied with intervening in the daily lives of miners and argued for measures to protect miners' safety—at least they did so rhetorically.[29] Two years after Mexican workers in Cananea struck for equal pay in the presence of US workers, the Díaz government secured an agreement among companies to "watch for the lives of miners," which recognized the "notorious danger of their occupation."[30] Yet the Cámara argued the agreements never "guaranteed the lives of miners in the absolute."[31] Nonetheless, mining inspectors attempted to implement punitive fines in order to limit worker deaths during the late Porfirian period as they outlined preventive measures to stop repeated accidents. Engineers emerged as the poster children for safety as the government emphasized their role in stopping accidents before they happened. While miners discussed "ills that happened with such frequency" in their tales of accidents and occupational harm, workplace statistics displayed more neutral faces to these occurrences. According to the *Boletín*, mining company supervisors suffered limited if any punitive damages, since the Cámara emphasized that "enough already exist[s]."[32] Even as mining companies increased the number of precautions to limit injury in print, few mining company owners exerted efforts to assure their implementation. While life and death in the mines depended on regulation enforcement, few engineers did anything to emphasize such measures underground. In the absence of local control, the federal government neglected to regulate safety issues and instead charged companies with emphasizing their own safety measures to guard against occupational hazards.[33] Mining engineers and inspectors therefore played a crucial role in keeping statistical tabs during visits to the mines.

MEDICO-LEGAL CONTEXTS OF OCCUPATIONAL SAFETY IN MINING

In late nineteenth-century Mexico miners received some degree of on-the-job protection from mining laws. These stipulated who could own mines, who could purchase property, which rights were associated with that purchase (water and/or mineral), and what measures would be taken to protect miners. The engineers who implemented occupational safety measures had in these laws a set of guidelines on what checks to perform for occupational safety. As part of their job, mining engineers were specifically named as part of this federal check. In the Mining Law of 1884, Título VI, Articles 121 and 122, mining engineers hired by local *diputaciones de minería* reviewed safety precautions.[34] But they did so only every two years.

Signed and passed by President Manuel González Flores (1880–1884), the Mining Law of 1884 served as the first step in opening Mexican lands to foreign extraction. While foreigners could not technically exploit mines, the 1884 law facilitated the broader exploitation of metal mines throughout the republic. In Título VI, Article 120, legislators stipulated the following basic requirements to protect miners: necessary ventilation by natural or artificial means, wide interior paths (the number of which was determined by the number of workers), interior walls fortified by wood or masonry to protect against cave-ins, and roads to and from the mine to be kept free from detritus and tailings. Most important, the law accorded miners the opportunity to demand the construction of fortifications for their safety. Miners, more than engineers, served as the primary observers of occupational safety, which I discuss further below in case studies.

In the subsequent 1892 mining law, legislators largely sidestepped the question of occupational health and safety but did emphasize the importance of ventilation; mine owners had the legal obligation to assure the ventilation of the mine. But this was not done for the workers' sake. As the law delineated in Article 12, section 12, the purpose of ventilation was to assure facile "communication" between laborers in the mine and the mine owners outside.[35] The 1892 law also lacked the "checklist" of safety measures outlined in the 1884 law. Without any clear guidelines, mining engineers and mine inspectors had no guidance to assure the basic safety measures of the previous legislation. As the 1892 law states in Article 22, "[Mine] owners shall be held responsible for accidents that happen in the mines because of bad work and shall compensate for the damages caused to other properties [in the same mine]." In short, mine owners compensated other mine owners for industrial accidents, and workers' compensation received only a glance.

As the Porfirian mining laws failed to emphasize occupational safety, injured or hurt workers had little legal backing to demand compensation until the Constitution of 1917. According to the Constitution, as outlined in Article 123, section XIV, employers had responsibility for labor accidents and occupational diseases in workers "contracted because of or in the performance of their work."[36] Therefore employers paid an "indemnification whether death or only temporary or permanent incapacity to work . . . resulted."[37] In addition, legislation included the Ley Federal de Trabajo, ratified in August 1931 (then later rewritten in 1970), which laid out the specific ailments associated with trades and crafts. However, the haphazard application of these protections led to limited improvement in occupational safety.

Because mines are inherently dangerous places and a variety of accidents, mishaps, and tragedies can occur, miners and other mine workers alike suffered a variety of occupational safety accidents while on the job. As such, the following sections include case studies of different accidents that took place in the mining industry in Zacatecas. They are organized by the cause of the accident and center workers' experiences to understand the challenges facing engineers in implementing occupational safety measures.

EXPLOSIVES

When municipal officials in the city of Zacatecas attempted to supervise the use of explosives in the silver mines surrounding their city, they discovered to their horror a slipshod, haphazard use of that technology. Because mines surrounded the city center of Zacatecas, municipal leaders were nervous about the active explosions regularly taking place in close proximity to the population. In many cases mining engineers did not calculate the amount of explosive material used in detonations; instead the job fell to mine workers who learned on the job. These workers employed an informal knowledge of explosives as well as trial and error in order to finalize the detonation.

Mine workers who used explosives also handled and stored explosive materials in a manner that did not support occupational safety. In June 1885 local political leader Atenógenes Llamas and his municipal engineer, Enrique Carrillo, circulated a notice to caution against poor housing of explosives in the San Tiburcio mine, of which they were scathingly critical.[38] Regardless of their efforts, local businessman Juan Petit petitioned the city in 1887 to store explosives on the south side of the Loma de San Fernando. Llamas rejected the proposal after careful discussion with "intelligent persons" and consideration for the safety of the city.[39] In a more sinister episode, mine operators needed to house their explosives properly to prevent the materials from falling into the hands of would-be rebels. Concerned citizen Federico Palmer alerted the city council of Zacatecas to a theft of explosives in October 1896. He wrote of how someone forced open the door to a storeroom near the La Victoria mine and stole two boxes of explosives.[40] To control these hazardous materials, Llamas and other politicians underscored the threat to human life caused by the day-to-day activities of mining. Because the city and the industry shared the same space, hazardous materials and tasks menaced not only miners but also the entire population. Mining engineers for both the city of Zacatecas and the mining companies faced increased pressure to achieve different goals in a municipality with historical and economic ties to mining. The engineers hired by the city attempted to limit locals' exposure to mining effluvia and other toxic material from the mines, whereas the mine's engineers maneuvered regulations in order to contain contaminants outside the mine, in addition to their tasks inside the mine.

INDUSTRIAL ACCIDENTS

Mining engineers also served as a first line of defense in the protection of workers' safety once they recognized how engineering practices affected miners' lives. Mining engineers warned the *Boletín* that mining detrimentally affected workers' health and advocated for the importance of studying and quantifying them to protect the public.[41] The *Boletín* authors in 1910 recognized their readers as primarily engineers or at least people involved in the mining trade; therefore they only lightly touched on occupational safety in a general discussion. Mining engineers primarily served to maintain and build structures underground to protect worker safety and unfortunately often fell victims to industrial accidents themselves.

In order to avoid paying compensation, mine supervisors blamed injured miners for their own recklessness, particularly if their bodies evinced signs of violence, and more so if family members filed a lawsuit against the mining company. In 1891 engineer Don Manuel Icaza accompanied Enrique Würst, the German administrator of a section of the famous El Bote mine, to inspect the division between two mines within the sprawling site.[42] Two wooden beams separated the Mina San Rafael and Mina Clérigos deep inside the mountain, signaling to municipal and mining officials the division between property lines. Icaza ignored Würst and his colleagues' warnings and walked the beams to test their sturdiness. As he climbed them, the beams quickly gave way, and he fell 30 meters (almost 100 feet), suffering a cranial fracture that caused his death in the dark depths of the mine. Consequently, municipal authorities Francisco Zarate and Luis Cordova raised an investigation to see which mining company should pay damages for the death of Icaza or to determine if his death involved foul play.[43] Lawyers and mining company representatives argued that the fall happened in a space that belonged to no one. Furthermore, Würst and his colleagues had warned Icaza not to venture onto the wooden beams as the bolts did not have a solid hold on the wood or the mining walls. His failure to heed the warnings, according to the officials, led to his death. Eventually, a judge cleared everyone involved in the case, citing Icaza's own poor judgment.

In the Icaza case engineers and doctors informed judicial decisions in a forensic capacity, examining both the structural collapse and the fatal injuries. However, the case revealed three narratives used by mining companies to escape claims and undermined Porfirian appearances to protect miners. First, Icaza's family sought to place legal responsibility onto the mining companies, despite the location of the accident between the two mines. They initially hesitated to name one or both of the companies as legally responsible and, in the process, illustrated the crowded nature of the interior of mountains. Layers upon layers of mines revealed a tangled network of ownership and property claims below the surface, blurring the lines of responsibility.

Second, engineers outlined safety measures to protect workers; whether mining companies implemented these precautions remained another matter. In addition to absolving the mining companies, engineers were not condemned because the victim received the blame for his own death through recklessness. In the legal case family members also raised the question of engineering responsibility. Why did the beams collapse and why was it not supported well? In an ironic twist, Icaza walked across beams that he himself had monitored and maintained, or rather had failed to do so. However, this raised the question: Did engineers in the late nineteenth century possess a responsibility for the greater good? In most cases, the Porfirian científicos argued in the affirmative and offered some shining exemplars of engineering professionals. In Zacatecas, Enrique Carrillo, an engineer for the city, stopped a mine from further exploiting underground water tied to a local pond from which many in the city drew water.[44] While he received no plaudits for his actions, he did confront his superiors on the municipal Mining Council for not anticipating the confrontation between the citizenry and industry.[45]

Last, the injuries suffered by Icaza signified a failure on multiple levels, including in engineering, which illustrated the voracious appetite of mining in placing extraction first rather than considering precautions for miners. At the worker level, many miners arrived at the mine ill prepared for heavy labor, having to provide their own hats, lanterns, and tools. Without uniformity in footwear, lighting, or tools, engineers attempting to standardize safety faced a chaotic variability in which to make improvements. Engineers also lacked the practical knowledge for implementing these safety precautions. Their education primarily emphasized the physics and mechanics of extraction but did not provide them with the practical wherewithal to apply these calculations to real life. Some engineers mismanaged operations and neglected regular upkeep of structures, letting them fall into disrepair, especially if the mine failed to produce valuable ore. With miners sidestepping minimal precautions and engineers in poor practice to apply safety standards, occupational safety was a shaky illusion during the Porfirian era despite the call to social responsibility.

TECHNOLOGICAL TRAGEDIES

Humans and nature during the Porfirian era were also becoming increasingly mechanized, which affected occupational safety at a critical moment in Mexico's extraction sector. Miners incorporated new tools as an extension of their labor and physical strength, while developments in engineering concentrated knowledge into an educated class of individuals. Historian and philosopher of technology Lewis Mumford argued that the shift from tools to machinery in mining during the nineteenth century fueled modern capitalism. Machinery prioritized the end product, leaving workers' skills with hand tools by the wayside.[46] Channeling Marxist ideas regarding labor and exploitation, Mumford argued that the mechanization of labor distanced workers from the satisfaction of having participated in production. Trapped in this dichotomy of mechanization and hand tools is the human body. Who bears the responsibility if workers are injured by machinery? How does the human body change in this shift from hand tools to machines? In short, mechanization became a means to an end. The end product stemmed from a capitalist push to produce as much as possible to meet the needs of the market. While humans remain fallible in their own work, the incorporation of machinery raised questions in the late nineteenth century about who carried the fault when accidents occurred. The humans? The machines? The engineers who designed the machines?

In addition to trying to avoid deadly decisions, workers suffered gruesome accidents because of machinery and the new technology. Félix Hernández worked in the Mina El Diamante in 1891, a pit belonging to the profitable Negociación Vetagrande company, nearly eight miles from the center of Zacatecas.[47] Owned by various individuals, the mine continuously produced silver and drew investors from Europe and the United States because of its reliability and its malleable walls that allowed for the construction and drilling of several branches off the larger mine. During one attempt to expand the mine, at 2:00 p.m. Hernández placed dynamite in holes bored through the rock to expose new mining ground.[48] He quickly exited the mine in order to let the fuse reach the charges. Growing impatient for the blast, he went to relight the fuse, thinking it had gone out. As he turned a corner in the tunnel, the explosion went off, and he received the brunt of the blast, which drove a drill auger into Hernández's chest. Coworkers quickly notified his cousin Pascual Hernández, a laborer in the same mine, who ran to the scene and dragged his cousin from the mine. Félix died soon after. Miners often experienced a variety of physical traumas because of the increased use of steam engines, dynamite, and machinery introduced by foreign investors. Distance often compounded the injuries, as most accidents happened far from hospitals and medical care. Félix Hernández's coworkers made an effort to take him to the hospital but did not arrive at Zacatecas from Vetagrande until 8:30 that night, over six hours after the explosion.

As noted by other contributors to this volume, new technology began arriving from Britain, the United States, and Germany during the late nineteenth century, which forced inexperienced handlers to manage machines with which they often had little experience. Steam-powered apparatuses, hoists, and pulverizers all changed the industrial landscape while exposing vulnerable workers to industrial accidents because of mismanagement, poor application, or lack of experience. Miners Berardo Carreón and Francisco Hernández both experienced traumatic injuries during an October 31, 1891, cave-in in the Los Tahures mine located outside Sauceda, Zacatecas.[49] As descriptions of the bodies and injuries detailed, Carreón bore the brunt of the rocks and died of his injuries. The rocks crushed his midsection, broke ribs on his left side, and smashed his left leg. The injuries did not kill him instantly; a coworker, Cresencio Ríos, guessed he died from blood loss as Carreón was dead when help arrived.[50] Hernández escaped with a broken leg and shattered pelvis, debilitating injuries in any era. As with the young man who died in the explosion, municipal authorities called for an investigation to determine who, if anyone, should be held responsible for Carreón's death. According to the court records associated with the case, a mining engineer had earlier declared the area unstable, and therefore the fault fell on no one.[51] After receiving all testimonies, municipal officials received permission by Eufemia Jaramillo, Carreón's young widow, to inter the body of her husband.

The gruesome and violent nature of many mining accidents even led many lawyers to file cases in the criminal records alongside accusations of murders and burglary. The Criminal Code of 1855, Article 253, required that all municipalities investigate accidents for foul play and murder. Likewise, lawyers and archivists printed in large letters on the cover page of the Carreón file: "Who is at fault for the death of Berardo Carreón and the injuries suffered by Francisco Hernández?" First, lawyers attempted to establish an investigation to determine if someone should be held responsible. Witnesses gave depositions regarding the events that led up to the mechanical or accidental trauma. Who accompanied the victim? Did anyone warn the victim? Did anyone harbor malice toward the victim? Second, if no one actively committed a crime, lawyers moved to hold the company culpable unless released by witnesses or family members who suggested that the tragedy was an accident, not the result of negligence. Often family members recognized in their depositions how the mechanics and physics of beams or machinery inadvertently caused the death of their sons or husbands, which apparently released the company of any charge of negligence. In recognizing its dangerous nature, many witnesses simply acknowledged deaths as part of the occupation.

CAUTIONARY TALES

As workplace safety drew on the expertise of engineers to protect workers, genuine precautions succumbed to legal wrangling and the influence of mining companies. Mining engineers oversaw workplace safety and contributed to workers' rights rhetoric. However, little was done to implement real, worthwhile measures in mining, in effect failing workers in the end. While the process of professionalization and education contributed to the slow application of protective measures, mining engineers found themselves trapped between assuring productivity and confirming the safety of workers. If the sacrifice of workers' lives through negligence was a necessary evil, then mining engineers proved to be a success. While Porfirian científicos encouraged engineers to protect the public with feats of design, they neglected the actual enforcement of policies and actions to benefit miners' health. Despite a flurry of legislation during the Mexican Revolution that deposed Porfirio Díaz, the federal government continued to allow companies to determine which precautions to prescribe in order to curb injuries or disease. Without proper regulation or accountability, miners had little recourse, despite the promises of the Constitution of 1917. Despite a rhetoric of health and improved societies, engineers failed to hold companies to account. They failed to preserve the Vital Machine of Mexico: its worker.

CHAPTER 4

ENGINEERING THE PORFIRIAN LANDSCAPE

Technology and Social Change in the Basin of Mexico, 1890–1911

JAMES A. GARZA

In early 1911 the noted Mexican engineer and architect José Ramón de Ibarrola delivered a paper at the Science and Arts Centennial Conference in Mexico City's National Preparatory School. The meeting, held under the auspices of the Mexican National Academy of Jurisprudence and Legislation, gathered the nation's leading scientific, civil, and professional societies to produce forty papers that highlighted the pinnacle of Porfirian-era achievements.

During his talk, Ibarrola, representing the country's engineering academy, divided Mexico's engineering history into four distinct phases: the pre-Hispanic era, the colonial period, the independence era up to 1876, and the Porfirian era, or Porfiriato. While Ibarrola noted each period's major accomplishments he took care to mention that before the Porfirian government the quality of engineering projects was poor, and therefore he had studied in the United States. This all changed when Díaz took power, the engineer proudly proclaimed, and Mexico entered a golden age of engineering accomplishments that brought peace and progress, but most importantly, order. As the conference proceeded, Ibarrola must have felt a certain amount of uneasiness as news of increasing unrest tied to Francisco Madero's revolutionary movement reached the capital. For Ibarrola, Mexico's development was not exemplified by rebellion but by engineering and technical achievements such as Mexico's expansive railroad networks that were the product of investment and planning. More than anything else for Ibarrola, it was the Porfirian Gran Canal del Desagüe that stood as the crown jewel of engineering triumphs.[1]

Developed by the Díaz administration in partnership with several foreign firms, the most important being the prominent British engineering company S. Pearson & Son, the Gran Canal del Desagüe was primarily designed to rid the capital of the floodwaters and sewage that periodically threatened lives and hygiene in the capital city. In actuality, a *desagüe* had existed in one form or another since the colonial era, a dream of viceroys and urban planners, but never fully realized because of a lack of funding, technology, political stability, or simply human will.[2] During the Porfiriato, government officials, investors, and engineers adjusted old plans and invested new capital to shape the project in the form of a modern tunnel, an excavated canal, and other components, the most impressive being the 48 km dredged channel that cut through the Valley of Mexico's northern half, intersecting pestilent Lake Texcoco, the lowest lake in terms of elevation and the recipient of *aguas negras* from the capital's ever-growing population.

For the project's planners, controlling the lake's level would be beneficial to the city's long-term urban health. Long blamed for the city's health woes, the lake not only functioned as an open sewer but as a focal point for criticism of a rural lacustrine-based lifestyle. Once completed, the Gran Canal would bring permanent environmental changes, inadvertently facilitated by modern engineering, to a region long denigrated by metropolitan elites as backward and primitive. Yet these engineered transformations introduced social anxieties and fears to communities beset with ecological challenges and, as a result, sought to confront and negotiate with the outside agents who brought the canal to their vicinity.

In many different ways engineering projects serve as a conduit for new technologies and often permanently alter the lives of citizens, many of them in rural zones. This was the story in late nineteenth-century Mexico when the rush to modernize included a surge in the importation of new technology and foreign advisers, the latter eager to capitalize on the Porfirian thirst for belonging to the congress of "advanced" nations.[3] While Sir Weetman Pearson played a central role as head of the firm, he sent personnel from Great Britain to operate the machinery, in particular two engineers: John Body and Thomas L. Walsh. Mexicans also played key parts in the complex operation. Luis Espinosa, the head of the Desagüe under Díaz and the lead engineer, drew valuable knowledge from personal interaction with the project's technical and social components. Other Mexican engineers gained similar experience. The tools, the equipment, the surveyor's report, the familiarity with the steam-powered dredge, and in some cases the response to rural petitions provided valuable guidance on not only modern engineering techniques, but the forging of state tactics that are often in the background as engineers interact with policy makers and bureaucrats.[4]

In that aspect the Gran Canal del Desagüe functioned as a formative chapter in the history of engineering in Mexico. Although built with foreign assistance, the project provided Mexican engineers with invaluable technical experience in hydroengineering and canal building through rough terrain. Advanced construction techniques involving steel bridges and aqueducts were utilized. The project also introduced new technologies such as Portland cement, employed for the first time in Mexico in the canal's infrastructure.[5] The canal also provided Mexican engineers with the experience of building infrastructure in a rural area and linking it to an urban center. The most significant accomplishment was the use of social engineering. Mexican engineers gained valuable experience in dealing with petitions from communities affected by the canal, a modernizing project that involved constant negotiation and confrontation.

This chapter traces the history of the Gran Canal through these important aspects. While the canal's engineering components allowed national engineers to gain valuable technical experience from foreign experts, Mexicans contributed their knowledge in return, especially about the terrain and local social conditions. The canal's financial and diplomatic costs were numerous and a complex issue in their own right. Up to 1900 (the year of its inauguration), the canal stood out as the preeminent public works project in an era when significant foreign investment drove Mexican development.[6]

This chapter also discusses how communities in the Valley of Mexico's northern half encountered and experienced the Gran Canal. For local residents, or *vecinos*, the canal disrupted local and regional trade routes and long-standing lacustrine economies based on rapidly disappearing freely available sources of water. By building the canal across lake beds, engineers accelerated the desiccation of the basin's remaining natural surface water. Moreover, engineers appropriated water to float dredge equipment, upsetting local ecologies. Residents fought back, using scripts of resistance and negotiation. For the region's vecinos the introduction of the canal signified disruption and change to a regional economy based on fishing and light agriculture and a territory that had long acted as a trade conduit for the mining and pulque haciendas immediately to the north and northwest. Ironically, the canal became a conduit for new engineering development in Mexico, acting as a training ground for engineers and bringing a new chapter in the regional integration of Mexico and the promotion of the country's engineering development.[7]

MACHINES AND LANDSCAPES

Some of the earliest ideas for draining the Basin of Mexico date from the colonial era, for instance, the sixteenth-century project known as the Desagüe de Huehuetoca, a venture that involved diverting the Cuautitlán River and tunneling into the hills in the basin's northern end.[8] The tunnel failed, resulting in an open trench. Although hampered by conflict and political turmoil, efforts continued in the postcolonial era. During the middle of the nineteenth century engineer Francisco de Garay came up with a plan that involved a main canal, a tunnel, and a system of tributary canals to connect Lakes Chalco and Xochimilco to a sewage system in the capital.

It was this project that eventually evolved into the Porfirian desagüe, but the political, financial, and technical pieces were not in place until the late 1880s, when the Díaz administration had improved its financial standing overseas and formed a semigovernmental unit, the Junta Directiva del Desagüe, to carry out the drainage plan. After a failed start, the junta contracted an American firm, Bucyrus, to work on the principal excavations of the Gran Canal using simple scoop dredges. They decided that a route from San Lazaro to Zumpango was the most feasible. Actual work began in 1888 while another firm, Read & Campbell, was contracted to work on and complete the tunnel. After some political and financial wrangling, the Mexican government and the junta ultimately decided to replace Bucyrus with the British firm S. Pearson & Son. Most likely the initial loan to begin the work on the canal influenced the decision. Made through a London financial house, it was "suggested" that the selection of a British company would be advantageous.[9]

Francisco de Garay's open sky route, adopted by Luis Espinosa, took the 48 km canal along a route that began in the San Lazaro district in northeast Mexico City and then northward between the low-lying Sierra de Guadalupe and Lake Texcoco, intersecting the eastern edges of the lake. It then shifted northwest, dissecting Lakes San Cristóbal and Xaltocan, both actually shallow lagoons, before it exited the basin in the Zumpango region, tunneling under the Cerro de Xalpa for 10 km before finally exiting near the municipality of Tequisquiac and eventually emptying into the Tula River.[10] De Garay, who had a career as a professor in the National School of Engineers and achieved international recognition in Paris and New York from their respective professional engineering societies, served as director of the desagüe in the Valley of Mexico before handing those duties to Espinosa, who had been named auxiliary engineer for the desagüe in 1871. Espinosa had also worked on mining ventures simultaneously and was aware of the daunting task ahead. Espinosa not only inherited de Garay's national and international experience but also the best blueprint so far

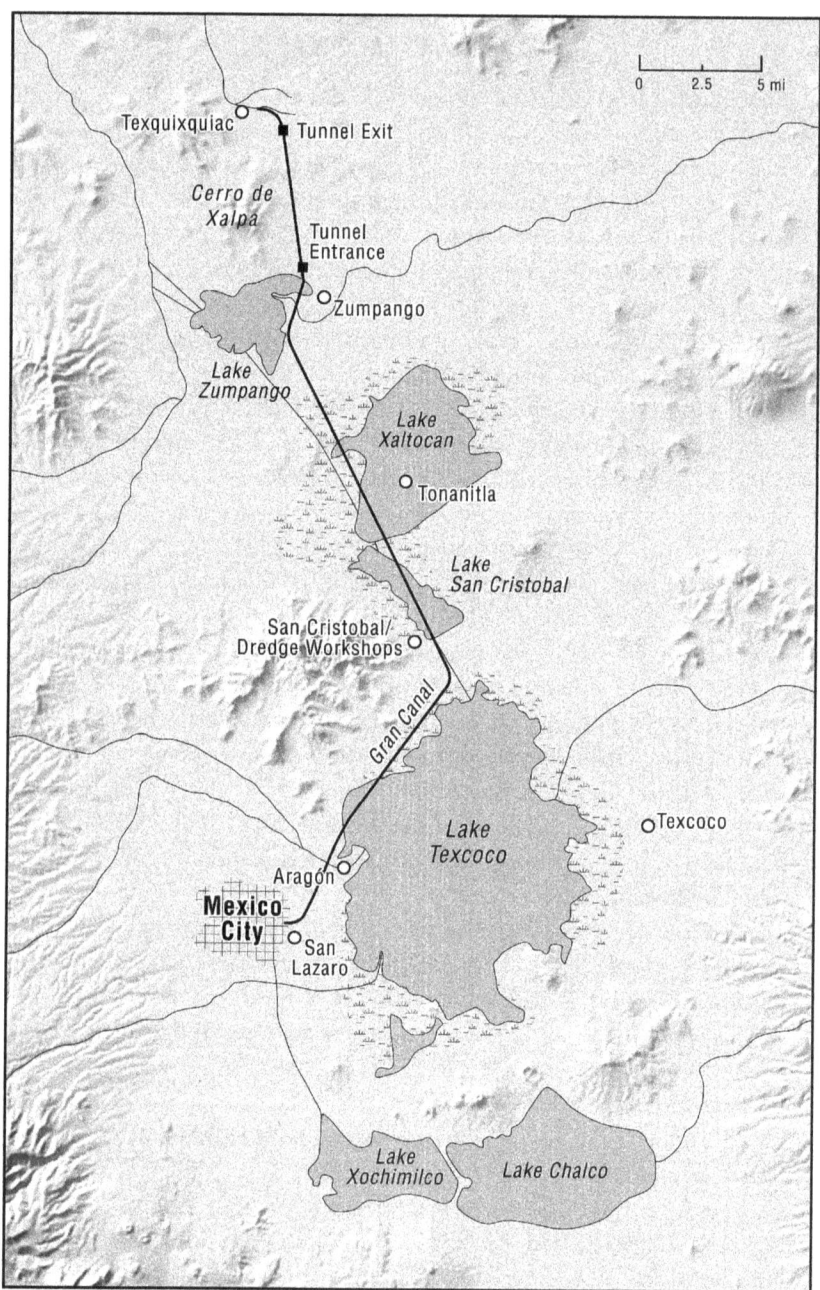

Fig. 4.1. Route of the Gran Canal del Desagüe, approx. 1900. Map by Erin Greb Cartography.

Gran Canal del Desagüe. Draga "Cuauhtemoc" kilometro 39. 1893.

Fig. 4.2. One of the Pearson dredges at work. *Source: Memoria histórica, técnica y administrativa de las obras del desagüe del valle de México, 1449–1900,* vol. 1 (Mexico: México, Tip. de la Oficina impresora de estampillas, 1902), between pages 456 and 457.

for solving the centuries-long problem. Bringing this experience to Pearson proved invaluable.[11]

As historian Paul Garner notes, Sir Weetman Pearson's transnational empire would eventually go on to make Mexico the center of its ambitious engineering enterprises. In the meantime, equipment flowed to Mexico to augment Bucyrus's simple scoop dredges. As John Body, Pearson's general manager in Mexico, wrote in his entry to the Institute of Civil Engineers in 1901, the Gran Canal's starting point in the olfactory-offending San Lazaro district, where the pumping station was located, was crucial to its success. From there the canal collected Mexico City's sewage as well as periodic floodwaters. A sewer system, installed in the best *colonias*, or neighborhoods, funneled the offending waste to the canal. It was fitting that the National Penitentiary would be located in the same vicinity as the starting point for the canal, visible reminders of the unwanted in Mexican society.[12]

Pearson's technology fit the bill. The firm brought in five advanced steam dredges, each employing the Couilor scoop system that consisted of two booms on either side that collected material scooped out from the front of the dredge. The booms then deposited the material on opposite sides of the canal bank. Work gangs subcontracted from the local population worked first to dig out the canal route to a depth of three meters. Groundwater flooding into the newly dug channel necessitated the use of centrifugal pumps, while the dredges were assembled on-site. It was a messy endeavor. Messr. Lobnitz & Co. of Renfrew, Scotland, constructed the dredges outside of Glasgow, took them apart, and shipped them to Veracruz on steamers, where they were then transported by rail and reconstructed on-site. Four of the dredges were identical in size while one was larger; all of them had two engines, with the main drive engine on each capable of 150 horsepower. Each one additionally had smaller engines to operate the winches and power pumps. The dredges were complex pieces of equipment for their time. They carried their own supply of coal and were crewed by fifty-nine personnel and one captain. The elevator dredges—named Carmen (after Díaz's wife), Annie (after Pearson's wife), Conchita, Lucy, and Cuauhtémoc—employed the same technology used in different excavation projects around the world, including the New York State Barge Canal, the Panama Canal, and the Suez Canal. After the project was over, the dredges were taken apart and presumably left in Mexico.[13]

It must have been quite a sight to see the dredges chugging along the Mexican landscape scooping out clay and tuff with their booms protruding like stiff arms depositing the soil mixture onto the canal banks to a maximum distance of 200 meters. Each dredge had the capacity to dig out 3,000–4,000 cubic meters daily, although this varied since some of the "soil" was actually a resistant tuff, or light volcanic rock, that was hard to excavate. The difficulties lay in making sure the excavated banks did not slide into the canal and the dredges did not run aground. As a result, the slopes had to be trimmed by hand. To fully operate, each dredge needed sufficient water in the channel to a depth of at least 15 meters. The most convenient sources were the ancient lakes of Texcoco, Zumpango, San Cristobal, and Xaltocan, whose basins were located adjacent to the canal and separated by earthen dams constructed by the dredge operations. As a result, the crews had to periodically break the dams or operate sluice gates.[14]

Engineers saw the difficult terrain north of the capital as a continual obstacle, and they employed additional technology in the form of a Lidgerwood Cableway to remove material that could not be reached by the dredges. The cableway consisted of two timber towers, 65 feet and 35 feet high, respectively, set on opposite sides of the canal and mounted on rails. Steel cables were run between the towers allowing for a system of buckets to be lowered into the canal bed to swoop in and scoop out material. While a small engine powered the lead tower, a small crew operated the trailing tower with a winch. Workers from villages throughout the valley provided the manual labor for adjusting the barrels and making sure the operation ran smoothly.[15]

One of the project's important goals involved making sure existing railroad, road, and river channels were preserved. Four aqueducts for existing rivers and canals were constructed (Canal Del Norte, Consulado, Unido, Guadalupe), and three bridges were built for important rail routes, including the Interoceanic. Officially funding was set aside for at least eight bridges, but the Junta Directiva deemed only the bridge linking to the Apam region and the Zumpango and Cuautitlán roads as important. They considered all the other connections as secondary and to be built only as needed, some in response to petitions and others after local demographics and needs were taken into account. Some crossings were never constructed. The junta deemed the Apam Bridge important because it was necessary for the pulque trade. The aqueduct bridges were constructed with a combination of steel and wood with masonry foundations, including the addition of Portland cement and stone.[16]

The canal's infrastructure also included sluice gates that disrupted water supplies and local ecologies. The project, an example of what historian Thomas D. Rogers calls "the transforming power of science and technology," changed the face of engineering in Mexico permanently, but not necessarily in a good way for local inhabitants.[17] For Mexico City the canal would bring benefits in terms of an outlet for sewage and floodwaters; for rural Mexicans in the canal's path, it would transform everyday life.

An analysis of the petitions launched by the various communities indicates the use of sophisticated scripts that point out economic losses and the awareness of ecological damage. Still, the petitions are guarded, with residents couching their language in terms officials would respond to favorably. While the responses are somewhat sympathetic, they also indicate a certain sense of indifference on the part of engineers and planners to local concerns. The common phrase "fuera de México todo es Cuautitlán" exemplified a disdain for the *provincias* that began directly outside the capital. While the region held little to attract Mexico City's cosmopolitan elite, the project's thirst for water to float the foreign machinery and the seemingly endless labor requirement to supply the work gangs became the only local resources of interest to the planners.

LACUSTRINE ENCOUNTERS

The region's alleged backward reputation certainly did not help it avoid condemnation. Several of these communities lay in the region known as the "salado," a plain that bordered Lake Texcoco which received its name from the high presence of evaporated salts along its shore, an environmental phenomenon that waxed and waned on an annual basis with the rainy season. In the 1840s a French citizen and traveler, Matheiu de Fossey, moved to Mexico City. He described the outlying region in unflattering terms as desolate and arid. He called the local population "Troglodytes" who lived in rudimentary dwellings and eked out a living in a treeless and ugly land.

Despite de Fossey's critical eyes, local residents made the best of it and mined the mineral salt *tequisquite*, a combination of chloride and sodium bicarbonate. They exported it to Mexico City where it had been used as a food seasoning since pre-Hispanic times and as an additive to pulque. In addition, rural dwellers practiced light agriculture and hunted waterfowl and fished on the lakes. They also harvested the local insects that infested the local shorelines and waters.[18]

Perhaps tequisquite's use in pulque added to its reputation as an Indigenous industry. Many urban developers perceived it as marginal to the modernizing economy. The valley's middle and upper classes certainly believed that rural products, such as the insects that were consumed as food, were unimportant and therefore did not care for the rural lifestyle. To add insult to injury, colonial-era drainage projects had begun the slow process of Lake Texcoco's ecological deterioration, but it was Mexico City's increasing population that contributed to Texcoco's reputation as a poisoned zone. In the last quarter of the nineteenth century, as planning for the Desagüe accelerated, city planners had increased their disdain for the ecological conditions produced in and around the lake.

In April 1878 a particularly dry period began, which produced an unbearable stench that permeated the capital. An official investigation concluded that the smell's source was the remains of insects and other organisms in the now-dry Texcoco, whose water levels were increasingly sensitive to inflow from rivers and rainfall. While a commission was formed to study the issue, the drought persisted throughout the month, alarming the population. On the evening of April 30, as Mexico City residents went to bed, a bright, white haze filled the sky and almost obscured the moon. The next morning *capitalinos* awoke to find the air loaded with dust pollution, a sign that a notorious *tolvanera*, or periodic dust storm that brought misery and illness, had suddenly appeared.

Francisco de Garay went to the lake the following day to investigate and discovered that Texcoco was completely desiccated. De Garay journeyed to the deepest part of the lake, a site marked by a stone cross, and found the location completely covered with a layer of tesquisquite. With that mystery apparently solved, the worst however was yet to come. Later that afternoon, a torrential storm descended upon the valley and flooded the city. The following months brought more heavy rains, proving the need for a drainage canal, a project that had been planned by de Garay's desagüe office but that would not be carried out until a decade later by his successor.[19]

Scientific validation of the lake's and the region's pollution led to more periodic investigations. In 1884 Antonio Peñafiel, a noted scientist and doctor, described an expedition he took part in to the increasingly pestilent Texcoco. His report only added to the lake's black legend, not that the inhabitants of Mexico City needed any extra olfactory confirmation. The Ministry of Development had commissioned Peñafiel to study Mexico City's water delivery system. Unfortunately for him, part of the study involved a field trip to the area where the San Lazaro Canal emptied into Texcoco, a location not known for its salubrious microclimate. Peñafiel, whose mind was already made up, blamed the lake for Mexico City's hygiene problems and labeled the body of water a "decomposing cadaver." Nevertheless Peñafiel and his crew pushed on, determined to take water samples for his study. As the excursion neared Texcoco on foot from the southeast, they noted the canal's fetid waters, contaminated with industrial waste, emptying into the lake. Then, to their surprise, they encountered a vast region of millions of dead snail larvae roasting in the heat of the day, giving off an exquisite stench that wafted in the still air and tortured their middle-class sensibilities. The exploratory party soon arrived at the shores of the lake and to their horror encountered what can only be described as a stretch of human waste covering the surface of the once proud lagoon. After taking their samples, Peñafiel and his companions must have been relieved to head back to more comfortable surroundings.[20]

Peñafiel concluded that short of avoiding dumping Mexico City's waste into the lake, the best solution was a drainage canal north of the city. Certainly Texcoco's ecological woes added to the urgency. By the late 1880s the Díaz administration had improved its financial standing to acquire enough funding to begin work on the canal. This came with a substantial price, however, as the region north of Mexico City was not lightly populated. As the canal dredging crews carried out their work, engineers encountered varied challenges along the way in the form of physical barriers such as rivers and communities whose way of life would be impacted. Since the pueblos depended on the scarce resources available to them, these challenges were intertwined.

Despite the potential effects on the local population, Pearson and the Junta Directiva signed a deal in the late 1880s. They began work on the dredging process in 1890 literally in the middle of the region from a base in the municipality of San Cristobal Ecatepec, roughly the midpoint of the planned canal route. The plan called for dredging to take place almost simultaneously at both the north and south ends of the route, although preliminary excavation by hand had previously occurred. Two or three dredges were to work on the northern end, in the Zumpango region, and one dredge would work in the south, near Mexico City and the villa of Guadalupe Hidalgo. Pearson also planned from the very beginning to build at least two temporary bridges and aqueducts for local and railroad access. Initially there existed the possibility that these bridges would swing to allow the dredges to pass. However, Luis Espinosa balked at the cost, especially at the idea of a swinging aqueduct that he presumed would be an engineering nightmare. The junta agreed, and the plans were shelved.[21]

With Pearson's idea of swinging bridges nixed and the exigencies of budget overruns surfacing, local residents began to complain when the canal cut off their access to local markets and to Mexico City. On February 28, 1891, the residents of the municipality of Santa Anna Nextlalpan sent a petition to the junta protesting that the canal threatened to cut the road giving them access to neighboring communities and the capital.[22] Although located in the Zumpango District at the northern end of the canal route, Santa Anna had been one of the first communities impacted by dredging activity. Census records indicated that in 1885 the district, located in the state of Mexico, held a population of approximately twenty-six thousand individuals. Nextlalpan was one of six municipalities in the district and contained four pueblos within its border, including the *cabecera* which confusingly held the same name as the municipality, and an additional five barrios and one hacienda. The entire municipality was situated in close proximity to Zumpango, one of the smaller lakes. A survey of the adjacent municipality of Zumpango, which gave its name to the entire district, revealed that the entire region relied on agricultural production of various cereal crops as well as fruits and vegetables, although water was scarce and of generally poor quality, with potability a rarity. Lake Zumpango was abundantly stocked with ducks, however, providing an alternate source of food, although seasonal variation in rainfall and the colonial-era Nochistingo cut into the mountains, deriving from the Huehuetoca Desagüe, made fishing unreliable. The lake, and the entire region, seemed to be in permanent ecological decline. Nevertheless, a local market on Friday attracted workers from the surrounding pueblos and haciendas, many of whom subsisted on the region's meager products. They haggled and worked together in a mix of Spanish, Otomi, and Nahuatl.[23]

Santa Anna Nextlalpan, despite being the municipal headquarters, was lightly populated and possessed no schools—the nearest educational facilities were located in Zumpango—and had no rail station. The canal proved to be an unwelcome intruder in the lives of the mostly Indigenous population. The vecinos cited not only the threat of their road to Mexico City being cut but a lack of access to neighboring districts. In his reply to their concerns, Luis Espinosa understood the economic consequences and proposed to the Junta Directiva that they order Pearson to construct a temporary passage over the excavation site where a centrifugal pump was located. Once the dredging was complete, a floating bridge could be constructed consisting of canoes or other rudimentary materials, but a swinging bridge was ruled out.[24]

Mounting tensions between Pearson and the junta may have also played a role in Nextlalpan. By 1891 the junta and Pearson were having difficulties caused by a variety of factors, including delays in the project's completion.[25] As quarrels mounted, local communities found themselves under increasing pressure as the canal's disruption of their trade routes altered daily life. In the small community of Tonanitla, situated on an island in the middle of Lake Xaltocan, residents petitioned the junta for a bridge in 1892, citing "damages." A delegation of ten vecinos wrote that while the junta had provided them with access to canoes, service was irregular and what they needed was a more permanent structure or, barring that, two community-owned canoes, one stationed on each side of the canal.[26] A resolution remained unclear. In any case, dredges working in the vicinity required water from the lake to float and operate, precipitating clashes between the residents of the community and dredge crews. At risk was Tonanitla's access to lake fish, including the *pescado blanco* imperiled by declining water levels.[27]

Vecinos, all men, protested to either the junta or Pearson utilizing the language of economic damage, whether from the loss of resources or the threat posed by the potential lack of access to local or regional markets. Romana Falcón, in her analysis of the art of the petition in the Mexican long nineteenth century, cites campesinos' defense of land rights, access to resources such as water and the right to use roads, especially in the central highlands, as frequent complaints in petitions. Although vecinos in Zumpango did not deploy the language of "time immemorial," their emphasis on the local economy in their relationship with the metropole dominated their petitions. At risk was something greater than a road to the local *tianguis*: a social compact that underlay campesinos' spatial and monetary relationship with the centers of power. The lakes served as a regional foundation of local capitalism, now being drained figuratively and literally as the local economy shifted toward a new orientation and the Porfirian administration placed a greater emphasis on public hygiene and the positivist slogan of order and progress.[28]

Perhaps aware of that, the residents of Tequisquiac, located on the other side of the Cerro de Xalpa, petitioned for relief directly from the administration, sending a letter to the Ministry of Communications and Public Works in Mexico City complaining that heavy rains had caused the Tequisquiac River to flood, cutting the community in half and isolating it from its regional market in the adjoining Mezquital Valley and from Mexico City. Specifically, the community noted that the canal's discharge, emptying from the tunnel under the Cerro, contributed to the river's increased carrying capacity and was making the town's inhabitants miserable. Couching their language in terms regime officials could understand, the vecinos cited commerce, agricultural, and industrial development as their primary concerns, noting that under the government of Porfirio Díaz Tequixquiac had recently enjoyed prosperity. However, the flooding threatened the community's livelihood, leading to drownings and even loss of access to the local school. The *ayuntamiento* asked for the construction of a bridge and cited the potential use of the desagüe railroad, the service line used to bring parts and equipment to canal dredging sites, as a future benefit.

The northern end of the canal route possessed vocal populations, but they did not monopolize complaints. Petitions from the southern end of the canal had also begun immediately after the dredging commenced. In the more centric municipality of Guadalupe Hidalgo, the small hamlet of San Juan de Aragón complained in 1891 that Pearson employees had diverted water from the Guadalupe River, depriving them of resources for agricultural use and to clean their clothes. Apart from this complaint, the villagers also noted that the canal had cut off access to Mexico City.[29] Isidro Díaz Lombardo, one of the senior Mexican engineers working on the canal and inspector of the overall project, responded dismissively that the water from the river was of little importance and would not be diverted after all. He added that the campesinos utilized a road that was not a real functioning route to the capital, just a rudimentary footpath that was frequently erased by the cyclical nature of the lakes and that there were alternate, more orderly routes to the capital, including the newly built road next to the canal. In doing so Díaz Lombardo dismissed the traditional routes long favored by local communities. The engineer also noted that the water would be utilized in the production of mortar by the contractors, so it had some value, but not in the way the villagers had envisioned.[30]

For Pearson diversions of streams and rivers were a common practice during the canal's construction since the planned route intersected several natural waterways. Since the dredges needed a certain level of water to operate properly, this process necessitated engineers shifting water from the lagoons, effectively taking a natural resource that vecinos needed for fishing or farming.

In 1888 the Ley Vias Generales de Comunicación (General Law on Communication Routes) federalized water sources in Mexico if a river, lake, or stream met one of three categories: it served as a dividing line between two states, was navigable, or had a significant value as a natural resource. While the law gave the federal government only jurisdiction, not property rights, over water, it nevertheless justified de facto government control over resources, a process that had been steadily growing in one form or another since the Reforma, as historian Raymond Craib states.[31] It is important to note, though, that local conditions varied from region to region. In some areas communities held considerable power over local water rights and would continue to do so even after the 1910 Revolution. However, in the case of the Texcoco region, environmental conditions and the encroaching presence of the Porfirian and later the postrevolutionary state diminished local water rights, even if regional communities negotiated against the onslaught of economic, political, and demographic changes.[32]

While Díaz Lombardo seemed to stick to the letter of the contract between Pearson and the Mexican government, Espinosa was more willing to entertain local petitions. This reflected some accommodation on the Junta Directiva's part, but it was limited; while petitions continued throughout the dredging phase, some of the biggest problems came from the project's financial and technical dimensions. Relations between Pearson and the junta deteriorated due in large part to the problems posed by the difficult terrain. Frequent landslides and clogs dogged the project, delaying completion, particularly in kilometers 0–9. As a result Sir Weetman Pearson traveled to Mexico in 1894 to renegotiate terms, and by December 1897 the entire project had been transferred to the junta for final completion, with gratitude given to Pearson for the firm's efforts. Despite the optimistic notices delivered by Díaz to Congress, technical problems persisted up until the inauguration date on March 17, 1900, the biggest headaches being landslides and inconvenient land shifts that jeopardized the projected connection of the canal to the capital's new sewage system.[33] Nevertheless, a lot of credit has to be given to Luis Espinosa for his efforts to "engineer" the canal's impact across the terrain. While the impact would be felt across numerous communities, accommodations were made if possible.

ASSESSING THE FUTURE

In 1894, as the Junta Directiva sorted through petitions, they received a letter from the *jefe politico* of (San Cristobal) Ecatepec requesting assistance to close the road named El Salado or La Mestla. The jefe noted that he had received recent reports of assaults and robberies on the road, as well as information that stolen cattle were being trafficked via the route. The road, mostly used for light commerce by a few *arrieros*, serviced small pueblos in the area, with little security. The jefe rationalized the closure, stating that the road would not be missed since it would increase security in the surrounding communities, benefit the republic, and most importantly, cancel the need for an additional bridge over the Gran Canal.[34]

Ecatepec's petition signaled the transition to a new phase in the region's economic and social development. The jefe politico opined that the old communication routes were unnecessary and not worth the effort, since unsavory elements would be less motivated to remain in the area. For their part junta officials, in their rationale for assigning bridges, deemed certain communities too small and instead concentrated their efforts on major crossings. They disdained traditional footpaths and cast the lakes as natural inconveniences and temporary resources to be used for development instead. The future was urban and concrete, not lacustrine. Rural Mexicans defended their traditional economic and social world, emphasized community needs, and sometimes utilized the phrase "order and progress," as in their petition to the jefe politico. It was a process of negotiation. Yet it was uneven since the Porfirian state, aided by foreign capital and technology, rearranged the landscape and granted them access to routes in the future.

That future was also uneven. When the canal was inaugurated in 1900, some people believed the salvation of Mexico City was at hand. While demographic growth and sinking ground levels ultimately minimized the canal's effectiveness, it proved useful until the middle of the 1920s, when the need for better pumping systems proved evident.[35] While the Gran Canal may not have solved Mexico City's long-term sanitation needs, it furthered the integration of the capital's hinterlands into the metropole. The canal's railroad provided another alternate route to the capital. As late as the 1930s the rail line's connections extended deep into the Mezquital Valley, allowing fresh produce to flow into Mexico City.[36]

The canal was an important milestone in the history of Mexican infrastructure engineering, linking rural and urban networks. While the canal's construction may not have left behind specific construction techniques necessary for Mexico's future in canal engineering, the nation's engineers nevertheless gained valuable work experience, for instance by working with Portland Cement.[37] Numerous Mexican engineers in addition to Espinosa and Lombardo, among them de Garay, Miguel Iglesias, Ricardo Orozco, Ángel Anguiano, and Fernández Leal, labored on the project. Roberto Gayol worked on the corresponding sewer systems, an overall part of the Desagüe.[38] Pearson's work also provided an opening its contracts on later projects, including developing Mexico's oil industry, another important training ground for Mexican specialists.[39]

The Gran Canal imparted invaluable experience to its Mexican engineers in another important aspect. While the British engineers may have proceeded on a business-as-usual basis, for Mexicans such as Luis Espinosa it was important to pay attention to the social aspects of the canal project. He was not an intermediary blindly enforcing top-down mandates; he was a mediator. Community complaints proved to have some merit and the junta did go to some length to accommodate local concerns. More important, the canal project gave engineers a taste of how extensively a modern project impacted rural communities.

Overall, Mexican engineers gained valuable experience working on a complex project with a lot of moving parts, sometimes quite literally. The canal's impact on the region's environment would be felt for decades to come.

Interestingly the Porfirian regime may have benefited the most, but not in the most apparent way. The Díaz government utilized the canal to boost its self-promotion at international conferences and expositions, framing it as a triumph of hygiene and engineering. At the Second Pan-American Congress in Mexico City in late 1901 and early 1902, delegates did not hear the petitions of those affected by the project and instead celebrated its recent completion and its promise to transform Mexico City into one of the great and healthy capitals of Latin America, an achievement for engineering order and progress throughout the Americas.[40]

CHAPTER 5

REVOLUTIONARY TECHNOSCIENCE

Science, Industry, Education, and the Mexican State, 1910–1946

JUAN JOSÉ SALDAÑA

The 1910 Mexican Revolution's demand for fair and democratic elections brought gradual expectations that went beyond electoral changes to include social and economic reforms. Within this context the Scientific Society José Antonio Alzate convened the first Mexican Scientific Congress from December 9–14, 1912. Thirty-six scientific and technical institutions participated in the exhibition. The congress sent a representative commission to meet with Vice President and Minister of Public Instruction José María Pino Suárez to express the consensus that scientists and technologists should participate in the country's necessary economic and social reforms.[1] But on that occasion neither the vice president nor President Francisco I. Madero (1911–1913), in his speech to inaugurate the congress, showed that science and technology in society were top priorities. The topics were put on the back burner.

Starting in 1913 the Constitutionalists, who took up the revolutionary mantle after the assassination of Madero, began to integrate a program of reforms connected to the demands of new social actors (peasants and workers). Constitutionalist officials hoped these reforms would help restore the legality interrupted by Madero's overthrow. At the opening of a 1914 revolutionary convention held in Mexico City, Venustiano Carranza, the first chief of the revolution, proposed to begin reforms even before the restoration of constitutional order.[2] This became an important point of dissension that divided the revolutionaries and threatened civil war. Despite this tension the provisional Constitutionalist government launched the reform program and proceeded to decree laws to resolve issues in the agricultural and labor sectors as well as to obtain greater control over natural resources. Thus, amid a military situation, the revolution's scientific and technological policy, still formally unpublished, began to be conceived and carried out to guide policy decisions. It was a policy in action in which engineers, doctors, architects, lawyers, agronomists, veterinarians, surveyors, geologists, climatologists, and other professionals were invited to participate in large numbers.[3]

Acting under this guidance, the Constitutionalist government replaced counterrevolutionary personnel and political opponents with people within the boundaries of Constitutionalist ideologies and reforms.[4] The change included a long-term scientific policy to make viable the special economic and political purposes of the revolution. It can be highlighted in three plans: First, a novel political, administrative logic was introduced by which preexisting scientific and technical institutions were transformed or eliminated. Second, existing or newly created institutions were to be organized in an innovative manner and with administrative and epistemic procedures designed to make education even more utilitarian in character and focused on the country's conditions. These factors were all deemed necessary to be able to contribute to the realization of the revolution's social project. Third, the institutions had a utilitarian orientation in the application of technical knowledge. Revolutionary officials and educators established more practical schools of engineering, medicine, industrial chemistry, agriculture, veterinary science, aviation, and others. These schools were aligned with the construction of aeronautics, armaments, tramways, and radio communication. They were also aligned with the legal, technical, and fiscal regulation of the oil industry that until then had largely been exempt from state supervision.

These measures reveal the presence of lucid minds committed to a revolution of national conscience in science and technology wrapped in the expressed need to turn to scientific and technical talent to build the new country.[5] In a sense these efforts stemmed first from political approaches dating back to Mexico's independence era, a longtime longing to govern and obtain the public good through knowledge and reason. The subsequent half-century struggle between liberals and conservatives in the field of education and the use of science remained present in later discussions about modernization and economic development. More directly, revolutionary political leaders and technocrats built on scientific and technical capital formed during Porfirio Díaz's government, capital that had to be taken out of the "box" and turned toward revolutionary change that would more broadly and rapidly address social needs.

The experience of governance with the utilization of technoscience between 1914 and 1917 fueled the conception and social organization of producing Mexico's wealth. State control of the country's natural resources and the education of Mexicans in terms of scientific and technical training became a fundamental part of the scaffolding that would build a new society sanctioned by a new constitution. The Constitution of 1917 was a remarkable political synthesis that provided the country a normative framework for its transformation and effective modernization. Together with a national scientific concourse, it provided a path for the future that would not be abandoned.

This chapter discusses the Mexican state's development through a scientific and technical framework, beginning with the immediate revolutionary era to the end of the postrevolutionary period in the 1940s. Political leaders saw value in developing Mexico's natural resources and began through direct state intervention to develop industrial, technical, and educational resources to increase the country's wealth and standing internationally.

SCIENCE IN ACTION

Thanks to the provisional government's successful scientific policy, it was possible to forge a new beginning in national science and politics because the 1917 Constitution assigned the state the power to intervene in strategic aspects of the economy and national life that intersected with science and technology. An example was the Ministry of Industry and Commerce, which was created in 1917 and whose chief engineer was Alberto J. Pani. He proceeded to reorganize several existing technical commissions, together with the former Institute of Geology, into a new unit: the Department of Exploration and Geographic Studies. Further, the National Geological Institute, under the supervision of petroleum engineer and geologist Ezequiel Ordoñez and with appropriate means (such as the first oil chemistry laboratory that existed in the country), became functional and was dedicated to researching oil fields, riverbeds, and other geologic formations. In 1916 the *Anales de Instituto Geológico Nacional* started publication and began to print the nation's new scientific policies. Finally, in 1918 oil exploration by Mexican experts began in great earnest, starting with the nation's Pacific states. In this manner the Mexican government directly entered and attempted to control oil production, thanks to a long-term policy that, while being in solidarity with the needs of the country, made use of its own institutions and talents in science and technology to realize national objectives. Before the Chamber of Deputies in 1917, Carranza, referred to the knowledge in oil drilling acquired by Mexicans in the following terms: "There is a great distance between the absolute ignorance that existed in the official spheres and public in general on the subject and the knowledge that already exists on this matter."[6]

Government involvement in private industry proved able to create benefits. For example, in 1915 the provisional government, in an effort to defend the demands of workers and assure tram service to the residents of Mexico City, intervened in the Streetcar Company of Mexico and did not return control of the firm to its foreign owners until 1919. This complex business used modern administrative and technical methods and was run with great success by the Constitutionalist-appointed auditors who possessed public administrative and army experience. What is noteworthy is that within the space of four years the company was able to function efficiently despite the effects of new national labor legislation, the revolutionary environment, and the constraining effects of World War I, the latter making it difficult to obtain imported spare parts as well as rolling stock and electrical equipment. Significantly, the auditors, as representatives of public government, gained valuable experience from running a modern private sector firm, specifically with its technological aspects. The company's main function of serving the public provided insights into everyday techniques, such as how to charge tickets and a myriad of other tasks. More important, the firm's administrative knowledge ensured its survival as well as the rights of its workers.[7]

The 1917 Constitution promoted innovative educational and technical measures to improve and increase food production, but changes had already been made to the nation's agricultural sector by the previous Agrarian Law of January 6, 1915, which established Constitutionalist agrarian reform policies. To carry out industrial animal husbandry, the government developed the Department of Aviculture, which introduced new methods for the breeding and raising of domesticated poultry. The organization also supported studies from the School of Agriculture that would examine animal diseases "that develop where there is an agglomeration of birds, causing hundreds of victims."[8] The following year, given the large attendance for these courses, additional classes in cuniculture, beekeeping, and the development of the milk industry were held. Carranza also agreed to create four experimental agricultural stations "of first class and fully equipped with modern tools and equipment" to carry out essential work.[9]

Clearly, the demand for engineering expertise led to increased organizational efforts. On June 8, 1918, engineers and prominent government officials formed the Center for Mexican Engineers, which counted more than 250 members at its inauguration, including engineer and Secretary of Industry and Commerce Alberto J. Pani, Secretary of Communications and Public Works Manuel Rodríguez Gutiérrez, and Undersecretary of Agriculture and Development Amado Aguirre. This organization of engineers, which also included architects, was the result of their mobilization of the transformative work of the revolution. In subsequent years the center continued to play an active role politically and epistemologically as it became a nationalistic scientific and technical trade organization that assisted in terms of applied knowledge, infrastructure, construction (roads and irrigation), aiding in the scientific organization of Mexico's postrevolutionary society.[10]

Even though engineering programs were not always adept at keeping up with the most current changes in mathematics and physics, the Center for Mexican Engineers played an important role in the modernization of engineering, specifically in teaching and practicing technical expertise. After the assassination of Carranza in 1920, the interim government appointed the old revolutionary intellectual José Vasconcelos as rector of the National University. Faced with the demand for professional engineers ready for the nation's new socioeconomic reality, Vasconcelos went to the center in 1921 and asked for advice on how to reform engineering studies.[11] As a result, the university initiated a process of reform and innovation in engineering and the sciences more broadly, an important development for a modern Mexico under the impetus of the revolution. This reform was urgent. In a short time significant highway (1925) and hydraulic (1926) projects were undertaken, improving the life of the nation.

Within the many powerful responses the center delivered to Vasconcelos, the reasons can be ascertained for the immediate success of Mexican civil engineering, especially in construction. After surveying its members, the center's president, Miguel Ángel de Quevedo, proposed among other reforms to continue with "the progressive and evolutionary turn" that the school had acquired since 1918. Quevedo wanted the "school [to] be in accordance with the special circumstances of our environment" and to give "new orientations of a practical nature in teaching plans and programs . . . practical studies, as corresponds to the application of sciences . . . with different teachers . . . [and] outside the school in construction camps."[12] He specifically called for competitively hiring the most qualified and suitable professors, making the courses of study shorter, creating terminal degrees for those who could not complete their entire engineering degrees (in effect training them as technical assistants), and eliminating industrial engineering from the degree program because of its poor performance and replacing it with mechanical-electrical studies.

Quevedo's proposals extended further still. He called for the creation of petroleum engineering studies because of its importance as a resource for Mexico. He introduced chemical analysis exercises in the study of building materials. Taking into account the increasing use of new forms of transportation such as cars and trucks, his recommendations also called for separating the studies of railroads from those of bridges and roads, given the importance the latter two had for communication with rural areas, especially with the growth of revolutionary agrarian redistribution programs.

Quevedo also recommended conferences on "forest vegetation" and courses on administrative law, political economy, and sociology for "the promotion and management of diverse companies."[13]

In an era when Mexico embarked on infrastructure development programs vital to its economic modernization, especially in the areas of agriculture, industry, and commerce (as exemplified by the national commissions on roads and irrigation) the Center for Mexican Engineers began—thanks to its political influence and its almost five-hundred-membership roll—to influence the government in hiring Mexican engineers for the design and execution of the nation's road and irrigation projects.

This shifted the Porfirian practice of hiring foreign engineers toward hiring more domestic ones for national projects. This by itself was a major step because, as Daniel Reséndiz Núñez affirms, it was a kind of "nationalization of civil engineering that made possible the development and subsequent self-sufficiency of this field."[14] To this effect the Center for Engineering announced a December 1925 accord reached by an assembly of its membership that called for a commission to formally petition President Plutarco Elías Calles (1924–1928) to take into account Mexican engineers' technical and administrative skills in roads, irrigation, and ports projects that had been undertaken or that were in the planning stages, and that the center intervene in the administration of the National School of Engineers so that it could influence the training of future engineers. For the center it was an essential national cause. Its members argued that the practice of handing out publicly funded engineering projects to foreigners, often with mediocre results, had been disastrous for the nation. They further argued that past government policies had reduced the country to an inferior and humiliating position similar to a scribe of "international capitalism" without voice or vote on problems of importance to Mexicans.

The rising influence of engineers associated with the center corresponded to transformations that were taking place at the same time on a global scale, changes that established engineers as some of the most decisive social influencers of this era in which social organization increasingly equated to technical administration. This concept, derived from technocratic theory or from the "social responsibility of engineers" that was expanding in both the capitalist world and in the Soviet Union, held that society's problems could be solved with scientific solutions and that engineering professionals possessed some of the most important knowledge and skills to carry them out. However, at the same time Mexican engineers cited as evidence nationalist political reasons for their claim that they would be the guides for uniting workers and political and business leaders for the love for the nation.[15]

Statements from the center signaled the framework for how technology could prosper in a country whose policies derived from the political principles of the Mexican Revolution. It intended to establish solutions to long-standing and complicated developmental issues that the nation was going to need to address. Its members strove to lessen social inequity while also keeping Mexico an active participant in global trade networks, professional organizations, and modern construction trends. This seesaw act had to be accomplished with growing but still limited numbers of scientific and technological experts. This epistemic-political negotiation was imperfect but did in many ways prove successful.

To appreciate the importance of this modernization project and the impact it had on the training of engineers, it should be noted that in just five years after the National Commission on Roads began its work, construction on the México–Laredo highway (1,240 km) was underway, with large sections completed and paved with asphalt (the entire highway was finished in 1936), while the Matamoros–Mazatlán highway (1,277 km) saw similar progress.[16] Other completed highway systems included the México–Guadalajara route (635 km), México–Acapulco (458 km), and México–Orizaba–Veracruz (463 km), among others. To many Mexicans these examples represented the progress accomplished between the government and the country's engineers without the intervention of foreigners, whose contracts were not renewed.

SCIENCE, RATIONALITY, AND THE MEXICAN STATE, 1930–1939

By the start of the new decade, the revolutionary Mexican state had achieved a relatively peaceful and stable political transition due to the development of new institutions and political practices put in place during the previous decade. Despite the stock market crash of 1929 and the ensuing Great Depression, Mexico had weathered reconstruction and the economic turmoil fairly well, thanks to commodities income from petroleum and silver. In 1933 the Society of Architects and Engineers of Mexico asserted, "With the resurgence of mining, there will be an improvement in the transportation industry and a collective improvement will come if fruits and raw materials are paid in gold and national manufacturing is developed with imported materials."[17] In the political realm, after the assassination of President-Elect Álvaro Obregón in 1928, a provisional president (Emilio Portes Gil) and two substitute presidents (Pascual Ortiz Rubio and Abelardo L. Rodríguez) finished out Obregón's term. In 1934 the presidential election led to the peaceful ascension of General Lázaro Cárdenas to the highest office of the republic, complete with a plan elaborated by the Partido Nacional Revolucionario (National Revolutionary Party, PNR) in 1933.

The Plan Sexenal, conceived by ex-president Calles, the *jefe maximo* of the revolution, was a "thorough program of action . . . based on calculation, statistics, [and] lessons of experience."[18] These rational political considerations that defined objectives and how to achieve them were a novelty in Mexico and marked a new era in which "the Revolution would be realized as a peaceful and creative process." The notion of an interventionist state that had its origins in the 1917 Constitution constituted the nucleus of the era's revolutionary doctrine in which "the state is an active agent of management and order of the vital phenomena of the nation"—an issue of great importance for the socioeconomic development of the nation in the first Sexenio and subsequent ones.[19] During the plan's introduction, numerous public, political, and military figures were present, among them those who had obtained technical and scientific training such as Alberto J. Pani, Juan de Dios Bojórquez, Gonzalo Bautista, and Luis Enrique Erro. The major social problem was, according to the Plan Sexenal, "related to the distribution of land and its best use according to the national interest." The plan also mentioned the livestock and forestry sectors as being national interests that required attention. To resolve the agrarian problem and develop agriculture, the plan presented the need for a scientific study focusing on the republic's agricultural possibilities with the "creation of institutes, laboratories, and experimental farms." Irrigation projects underway were seen as "a necessary complement" to the development of agriculture, and for this reason the plan indicated that these projects should increase in number and coverage over the entire country. In similar terms, the construction of roads and airports were planned as "necessities most urgent."[20]

Other relevant government-related themes demanded scientific and technical investigations for the development of electricity, industry, public works, and public health. In terms of electricity, the report indicated that "the government procured a national system for the generation, transmission, and distribution of energy . . . to achieve the systematic supply of (electricity) throughout the nation." Concerning industry, the document stated that it was necessary to "stimulate the creation of new industries that would substitute domestic products for imported ones, or would take advantage of unused or underutilized resources," such as "maintain permanent exploration and experimentation services to support producers with all available technical assistance."[21] In the case of public health, above all else with the case of tropical diseases, the plan recommended the creation of an institute.[22]

The Plan Sexenal also called for primary and secondary education that emphasized "a scientific orientation" that would form part of the proposed "socialist" reform of Article 3 of the Constitution. Likewise, agricultural education would be prioritized in the form of hiring "trained technicians" to teach theory and practice to bring attention to agricultural issues. As a result technical training would have preference over a university education. The plan's provisions over the themes of public administration, science, and technology, designed to provide for the rational achievement of political objectives, constituted an important declaration. They reinforced a growing link between the two decades of cooperative existence between the revolutionary government and the field of science. State intervention in different aspects of national life was reaffirmed as official interest in creating research institutions increased. It was a clear effort to augment the presence of scientific research in all aspects throughout the country—an important step in the consolidation of science in the nation and its relation to the state.

The Plan Sexenal became a guide for all government and scientific activities. Testimonies exist that illustrate how the development and practical utilization of scientific activities under the plan were carried out. One of these belongs to Enrique Beltrán, who formed part of a group of Mexican scientists who had studied outside the country, in his case with a scholarship from the Guggenheim Foundation in 1931. He returned to Mexico with a brand-new PhD in zoology in 1933, becoming the first PhD-holding biologist in Mexico.[23] At this point, Cárdenas's presidential campaign was underway. So too was the creation of the Plan Sexenal in which Beltrán's friends—the economist Mario Sousa and engineer Alfonso M. Jaimes— participated. Both Sousa and Jaimes invited Beltrán to reform the Ministry of Agriculture. Beltrán intervened to propose the creation of a biological research establishment to be managed by the ministry. Outgoing President Rodriguez accepted his ideas on January 1, 1934, and the organization began operating in Mexico City as the Biotechnical Institute, with Beltrán not only as director but also as head of the hydrobiology section, itself a new discipline he introduced.

Beltrán conceived of the Biotechnical Institute following the regime's radical yet rationalist policies, a plan that correlated with his own ideologies. The institute's fundamental objective was to realize applied biological research to achieve concrete results for the social revolution. In the institution's statement of philosophy, he wrote that it is necessary that "the concept of UTILITY, derived from the possibility of immediate application, presides over all the work of the establishment." Research tasks would be defined as "indispensable to our research into our natural, animal, and agricultural resources," of which "immediate application" signified the use of rigorous scientific awareness of the environment.[24]

However, being a research center under the postulates of the Mexican Revolution, where nationalism prevailed, the institute's work also acted as an arm of revolutionary propaganda. It was necessary to show that the revolution "was not an abstract and lifeless postulate, nor a simple political change in the administration, but rather an organized effort to help the proletariat of the city and the countryside improve their living conditions. . . . Science, with a socialist criterion, has ceased to be a simple speculative task, becoming a servant to the needs of the community." In the same way, Beltrán thought, science would demonstrate how to improve productive activities in the field and convince campesinos "that the laboratory and the field are more useful than the Church, and that results are more effective in resorting to science in his daily life than appealing to a hypothetical God who is a product of his ignorance."[25]

Guided by these ideas, Beltrán organized the institute and hired the technical and scientific personnel, a task not without difficulty because of the absence of specialized personnel in the country. The institute had central laboratories (physics, chemistry, botany), offices of meteorology and agrology, and sections for plant genetics, animal genetics, marine biology, animal health, and plant health.

One aspect that stands out is that the institution's organization was based on objectives selected with a criterion of seriation:

> The relative economic importance of the natural products under investigation has been taken into account. Likewise, attempts have been made to order the studies in such a way that those practiced in one year serve as a precedent to those carried out in the next. Also examined are ways to arrange the studies so that technical material can be distributed to the various departments of the Agricultural Development Directorate at the executive level in an order necessary for work distribution. At the end of the Sexenio there will be, as far as economic conditions allow, a fairly acceptable knowledge of our natural wealth that today is almost totally ignored.[26]

As a result, the period 1934–1939 gained importance as a crucial time in Beltran's overall plan.

Counting on the support of Cárdenas, Beltrán designed and put into place plans for the creation of a limnological station on Lake Pátzcuaro as part of the institute. Its objective was to develop fisheries and conduct hydrobiological research.[27] Likewise, Beltrán commissioned a study titled *List of Mexican Fishes* in addition to other hydrobiological works for various lakes and lagoons in the country.[28] In its first year the institute initiated pioneering investigations in agricultural meteorology (related to microclimate studies) and plant genetics, specifically studying different varieties of maize and soybean seeds.

At its inception the institute lacked sufficient competent personnel and technical and economic resources. In addition, there were bureaucratic obstacles, since upper-level officials in the Ministry of Agriculture did not understand the role of a biological research facility.[29] A change of leadership in the ministry early into the Cárdenas administration motivated Beltrán to resign only a year after the Biotechnical Institute started its activities. The institute maintained its activities, although it changed its orientation in comparison to its original mission, moving toward veterinary medicine (with contributions to animal genetics) until 1940, when it was transformed into the Livestock Research Institute.

As laid out in the Plan Sexenal in 1939, the Department of Public Health put into operation the Institute of Health and Tropical Diseases. Planned since 1935, the institute, housed in a new building, was headed by medical-hygienist Manuel Martínez Báez and was endowed for research. Per the institute's regimen and with modern medical objectives, Martínez Báez held a nationalistic scientific philosophy: "Today we seek the resolution of our health problems not by the blind application of elementary norms taken from textbooks or by the servile copying of methods taken from other countries, but by the correct investigation of our own situation, our own particular needs, and our own resources."[30] It also had its own scientific publication, a favorable work environment, a group of magnificent professional researchers (among whom were Doctors Miguel Bustamante, Eliseo Ramírez, Geraldo Varela, Luis Vargas, Luis Mazzoti, Felipe Rulfo, and Enrique Beltrán) that made this institute the best in the country.[31] It should be mentioned that the institute continues to exist at the time of publication, albeit with additional functions and under another name, the Instituto Nacional de Diagnóstico y Referencia Epidemiológicos.

During the 1930s other events, also of an epistemic political nature, occurred concomitantly, leading to the founding of specialized institutions and to training reforms for the country's engineers, many of whom, encouraged by the engineer Sotero Prieto, went to study in US universities with scholarships from the Guggenheim and Rockefeller Foundations. Upon their return, they taught new concepts in mathematics and physical sciences in the National School of Engineering and later in the Faculty of Science that they helped create. Such was the case with Alfonso Nápoles Gándara who in 1932 introduced courses in mathematical analysis and differential geometry, topics that he had studied at Princeton. Moreover, due to the professional links established by these students, foreign faculty came to Mexico to teach and deliver papers at conferences. All of this academic activity produced an increase in the knowledge, training, and growth of a community that would be particularly dynamic in the promotion and institutionalization of physics and mathematics and their applications to engineering, which were in great demand in infrastructure construction.

On January 21, 1935, a new faculty of physical sciences and mathematics was constituted at the National Autonomous University of Mexico (UNAM) consisting of three previously separate but interrelated entities: the Department of Physical and Mathematical Sciences, the National School of Engineers, and the National School of Chemical Sciences. Ignacio Avilés served as its dean, with engineers Valentín Gama, Basilio Romo, and Ricardo Monges López leading the Departments of Exact Sciences, Physical Sciences, and Engineering, respectively. This new university institution had grown out of confrontations with state officials and out of an agreement with the University Council.

It was the first successful result in the country to coordinate engineering with other scientific disciplines, with updated plans and programs of study prepared according to precise academic objectives to train engineers and scientists. In correspondence with this plan, new careers in related fields were planned, some of which the Center for Engineers had already proposed, such as petroleum engineer, mechanical electrician, chemical engineer, and pharmaceutical chemist-biologist. These were added to the existing and reformed career options in civil engineering, chemical engineering, municipal engineering, and sanitary engineering.[32] Together with graduates from other institutions such as the National Polytechnic Institute (IPN), established in 1936, UNAM's graduates went on to form a technological contingent that promoted and carried out the country's material transformation.

The members of this institute formed with independently functioning schools and newly created ones to teach middle school, high school, and higher education on a technological basis while also carrying out scientific research at the National School of Biological Science. At the upper educational tiers, it created schools of mechanical and electrical engineering, extractive industries, engineering and architecture, physics and mathematics, chemistry, and others.[33] The need to train engineers in those years was necessitated by the 1938 oil expropriation, since it was necessary to replace the foreign technical personnel who worked for the expropriated companies and keep the industry running. Another important government unit that had been established in 1937 was the Federal Electricity Commission, which would become decisive in promoting electrical engineering and electrification in the country. Only with these decisive moves was the technological community able to meet the demands made of it at that time and contribute to the unprecedented development of the country in the following decades.

Additionally, the first modern scientific research institutes in the country were formed during the late 1920s and 1930s, such as those established by the UNAM in the areas of biology (1929), geology (1929), and later physics (1938). These establishments came to join the other research centers connected to national irrigation and road commissions, such as the experimental engineering and materials laboratories. These programs also interacted with the newly created Petroleos Mexicanos (PETROMEX, 1934), the Ministry of Communications and Public Works, and the National Railways. It should be noted that at this time researchers in the Experimental Engineering Laboratories applied for the first time experimental and analytical methods to innovate and solve civil engineering problems related to the field of hydraulics and soil mechanics and structures, both of which would have enormous implications in the coming decades regarding the sinking of Mexico City's buildings because of groundwater extraction and to the danger posed by earthquakes.

The Mexican state, as manager of the economic and social life of the country, followed an interventionist strategy and, as a result, recognized the importance of creating financial institutions to support development. This strategy contained objectives that required technical assistance, facilitated in 1933 by the creation of the Banco Hipotecario de Obras Públicas (National Public Works Bank) to finance urban infrastructure as well as the other public works carried out by the irrigation and roads commissions in the 1930s. The creation of the Nacional Financiera (National Finance, 1934), aimed at supporting the country's industrial development, also contributed to the modernization process.

One of the crucial commissions during the time was the National Irrigation Commission, an integral part of the land reform process. Beginning in 1925 it carried out, often without the assistance of foreign contractor companies, a large number of projects such as the El Palmito, El Azúcar, and La Angostura Reservoirs and other small and medium irrigation projects.[34] This required the specialization of its engineers in the various branches of irrigation engineering, creating what was called the "Mexican Practice of Irrigation Engineering" that would end up being exported to various Latin American countries, such as Bolivia and others in Central America which lacked expertise in this technique. In this way the studies, designs, specifications, materials control, and construction of the hydraulic works were carried out with a mode of study and execution appropriate to Mexico, leading to the construction of irrigation works of low cost and high efficiency. The adaptation of some natural structures and waterfalls for the generation of electricity was very useful for the industrial development of new cities and population centers in various parts of the republic. By 1940, thanks to the projects undertaken during the Cardenista Sexenio, it was estimated that a million hectares came under irrigation out of a total land area, estimated by various agricultural studies, of 2,987,500 hectares. According to the National Commission of Agriculture's 1940 *Informe*, the irrigation measures "significantly quadrupled the productive agricultural area in Mexico and consequently its food resources."[35]

The Irrigation Commission's experimental engineering laboratories were an important component of its mission, with its laboratories conducting studies on major projects, that produced positive results and contributed to the formation of Mexican specialists in hydraulic models and soil mechanics. Interaction with foreign technical consultants, as well as the dissemination of existing specialized literature, especially studies considered of great interest and utility, was important for this "live" training of specialists. This literature, in the form of "technical memorandums," included materials translated mainly from English and German and texts from some commission engineers; original brochures were published for those who had done special work. It is important to take into account that the specialized professional activity related to the construction of irrigation public works comprises a large number of the branches of civil engineering and the sciences (topography, hydrology, geology, structural design, chemistry, material physics, modeling, etc.), making it reasonable to imagine the magnitude of the task carried out and the knowledge invested by Mexican engineers in the execution of what was considered at the time "among the greatest works in construction in the world."[36]

The economic recovery that began in 1932 continued throughout the following six years with land distribution being a factor, reaching 18 million hectares among 800,000 *ejidatarios*, highlighting the fact that "half of the irrigated area of the country, the best quality, passed into ejido hands."[37] To this was added the network of highways and roads that were built and allowed the integration of the country and its economy. As a consequence agricultural production and exports increased. Apart from the economic factor, industrial development took place after 1933 at a high level because of the rise of oil and silver prices, both of which Mexico exported. Conveniently, the country was able to increase "demand and import capacity, which allowed the purchase of foreign raw materials which, given idle capacity, led to the rapid resumption of production."[38] Concurrent factors were also the development of a communications network that allowed industrial and urban centers to be linked (with reduced transportation prices), contributing to the growth of the domestic market and the ability of new industries to emerge and old ones, such as steel, cement, and cigarettes, to be reactivated. Their profits were then reinvested, leading to capital accumulation and a further increase in profits, resulting in industrial investment and, according to economic historian Stephen Haber, "manufacturing for the first time took the lead in the economy."[39]

According to what economist Enrique Cárdenas has established, "about 37% of the growth in industrial demand observed in the decade came from import substitution," which meant a rapid growth in demand and industrial production.[40] The government's adoption of monetary expansion policy, which increased public spending, was another factor for the growth of internal demand.

It should be noted that the industrial process that took place in Mexico during the 1930s coincided with the global modernization of engineering that occurred as a consequence of the Second Industrial Revolution. An example of this was the emergence of the scientific administration of labor and industrial processes, the engineering of production methods and other procedures to reduce costs, reduce waste, select and train workers, among other benefits. The industry that was "reactivated" in Mexico, or the new ones that emerged because of these economic impulses incorporated new technological advances. Thus modern technical education in institutions such as UNAM, IPN, technical schools such as Alvaro Obregón in Monterrey, and others had as a curriculum for their students' new engineering techniques. Many specialized fields of modern Mexican engineering, such as industrial, civil, electrical, mechanical, metallurgical, agronomic, petroleum, chemical, and others, can trace their development in the country to this decade.[41]

As illustrated above the Sexenio that began in 1933 introduced to the country scientific and technical plans that carried out Mexico's social and material progress. Within the framework of the country's social revolution and especially in the crucial decade of the 1930s, essential areas of daily life such as agriculture, health, public works, and industrial activity were successfully served thanks to science and national technology that were at the same time the cause and effect of the resulting progress.

INDUSTRIALIZATION AND SCIENCE IN MEXICO, 1941–1952

Handing over the office of the presidency in December 1940 to President-Elect General Manuel Ávila Camacho, Lázaro Cárdenas could boast of many achievements in the material, economic, and social progress of the nation: modernizing agricultural production, a growing economy with an advancing industrial sector, a strengthened and organized labor sector. But he also handed off a divided society. Various social sectors had resented ideological radicalism and Cardenista statism, and they had even managed to form an electoral political opposition that had significant support from the popular classes and the business sector, advancing Juan Andreu Almazán as candidate for the presidency. The official party and powerful national groups, recognizing dissatisfaction, drove politics in general toward a point of convergence with revolutionary politics and the achievements of the previous administration, resulting in a conciliatory strategy and the appointment of Ávila Camacho. While the general had not been ideologically identified with the Cárdenas government, Ávila Camacho attracted the majority vote in a highly contested election.

The Partido de la Revolución Mexicana (PRM) announced this change of strategy in the "Second Sexennial Plan, 1940–1946."[42] This plan recognized the need to incorporate other social sectors in an inclusive policy: "The revolutionary movement has reached a point where, although it has a very broad task to fulfill in the future, it no longer fears for its fundamental conquests," which authorized a call to "all forces that exist in the Nation" to ensure the development of its economy. The business sector was directly alluded to: "Guaranteed, as they are, labor rights, private initiative will find neither obstacles nor hostilities in the Six Year Plan."[43] On November 3, 1939, in his speech upon taking office, Ávila Camacho stated that in his government "new teachings and standards that time is perfecting and incorporating as the doctrine of the Revolution" would be important, as would the recognition of the division present in society that demanded another type of politics and incorporation of new social actors for which, he concluded, the government will not be of one party but of the entire society, "since a people is not a heterogeneous group of classes, each one bitterly defending its interests, but a great historical unity."[44] This new, multiclass nationalist political strategy was precisely what the party and its mass organizations presented in the second Sexenio and from which the government's ideology and political criteria emerged.[45]

About agriculture the plan stated that the government should "ensure As a result, industrialization based on science and technology became the dominant theme in the second Sexenio. The plan focused on development, with headings such as "Agrarian Redistribution and Agricultural Production," "Industrial Economy and Trade," "Mining," "Petroleum," "Industrial Electrical Production," "Communications and Public Works," "Work and Social Security (Pensions)," "Public Health," and "National Defense," all indicating various components of a technical and scientific nature, now deemed indispensable for the efficient governance of the country. As in the first plan it sought to infuse rationality into politics by establishing "national" objectives that scientists and technicians helped develop. Human resources and scientific potential were emphasized.

About agriculture the plan stated that the government should "ensure that the agrarian reform is carried out integrally in the shortest possible time." The plan also established that the reconstitution of *latifundismo* should be prevented and that the production of the ejido become the basis of the agricultural economy. But among other development measures (such as agricultural and ejido credit) the following objective was laid out: "The industrialization . . . of agricultural, livestock, forestry, and hunting and fishing products will be stimulated" and "the production of auxiliary industries of agriculture, such as machinery and agricultural tools, fertilizers, substances to combat pests, products for veterinary use, etc." To accomplish this, new vocational schools in agriculture and ejido administration were to be created to standardize agricultural curricula and degrees so they could all be "declared professions of the State." The plan further called for "the establishment and coordination of all scientific and technological research institutes, as well as statistical elaboration, for the dual function of providing, on the one hand, a technical basis for production, and on the other hand, to spread useful knowledge among producers."[46]

With regard to the economy and industry, the plan considered that a displacement "in the center of gravity of the economy" of nonrenewable resources should be carried out in favor of those that are renewable. Several measures were proposed, including increasing the electricity supply to provide more power to industries, improving the transportation system for the distribution and circulation of goods in the national territory, carrying out an inventory of natural resources, training industrial technicians, and establishing an office that would concentrate and systematize information from scientific and technological research institutes. The plan called for government action to "create and acquire permanent means of production and to provide the country with the mechanical work equipment necessary for its economic development . . . and to promote the establishment of those industrial plants that complement the extractive process." Mexico produced more energy, though most of it remained nonrenewable.[47]

The plan also stated that the nationalization of the oil industry would be total and final. It called for the industry to be technically organized and coordinated with the rest of the country's sectors, especially with the chemical, electrical, and war industries to establish conditions for reciprocal support and promotion.[48] The plan also called for the development of more fertilizer, iron and steel, cement, paper, and cellulose industries, and also emphasized the creation of new types of manufacturing plants to replace imported industrial products. Overall the plan emphasized the promotion of heavy industry and industrialists keeping up with the latest advances by replacing aging machinery in their facilities.

A section of the plan also outlined objectives for communications and public works, a vital component for development. The plan called for obtaining access to natural resources and establishing and improving circulation and distribution of products obtained from these areas. The merchant marine was also noted, as was the development of aviation applications such as the manufacture of airplanes and engines. In addition, it called for the preservation of existing radio communications, railways, ports and docks, as well as the completion of those projects already underway and the construction of new facilities with improved efficiency and technical accuracy.

One important component was education. The plan specified that the state "define exactly the ideological and pedagogical orientation of the Constitution's Article 3," a provision originating from the controversy, confusion, and nonconformity the Cardenista government caused when it modified the article in a sense to convert teaching to a "socialist" orientation. The plan worked to rid the government of "useless" establishments, showing an emphasis on utility to be provided by more efficient technocrats. The plan also called for the creation of cultural institutions and professional training for technicians, professionals, and other researchers.[49]

As can be seen from these examples, the PRM's Second Plan provided for both the government and Mexican society a program of political action appropriate to the transition Mexico was experiencing from an agrarian society to a modern, urban, and industrial one. This made it necessary to introduce new objectives concerning the previous Sexenio since the situation was very different from the one that existed in 1933. Essentially, to achieve social balance and maintain power, it required "collecting" claims of those who protested against state interventionism and the government's intense agrarian and labor policies. In the plan's new policy, formed with the participation of workers' organizations, the unusual call was made for private initiative to intervene as an important actor in the construction of a modern industrial economy. The importance given to industrialization and infrastructure in the government's plans was dominant. The looming issue of import substitution industrialization to strengthen national industry also became important.

Regarding work and its relation to the economic system, the new policy called for the government to create new possibilities and transform the existing system to avoid social injustice, economic disorder, and improve wealth distribution.[50] This important aspect rounded out the plan's modernization and economic goals. The purpose of the economy, the plan noted, was to promote and, in most cases, create Mexican industry that could cease to be protected by tariffs and subsidies and become capable of competing internationally, modernizing its equipment and methodology, and acting in concert with social justice.[51] These were, of course, long-term goals that would be carried out by multiple governments.[52] The same goals were relevant for agricultural production; it was to be modernized into a series of industries for the production of food and raw materials.

As a result of the incorporation of engineers and scientists into government structure, this modernizing policy had already led in the 1920s and 1930s to the rise of technocracy and technoscience in Mexico. However, in the 1940s the role of technocracy became crucial for the state. Since modernization implied that the generation and distribution of wealth were left in the hands of technicians, the design of social and developmental policies would correspond to them. It was understood that the construction of public infrastructure and industrialization were becoming central in society and that the relevance of science and technology were increasingly essential to economic production. Further, the effects of modernization in society (in health, education, food production, communications, social services, etc.) as well as in the environment (rational management of natural resources, environmental protection, and nature conservation) were increasing. Consequently both the governments of Ávila Camacho and Miguel Alemán (1946–1952) directed initiatives toward these aspects of economic and social development, especially for the strengthening of national science and technology and in some cases for private initiative as well.

A few months after the new government began, on May 13, 1941, the country's industrialization project was launched with the Transformation Industries Law, legislation that established tax exemptions for five years (in 1946 they would be extended to ten) for new or necessary industries. The law also established a mechanism of permits for the importation of raw materials, machinery, and equipment that were not produced in the country, thereby promoting national industrial development.[53]

In 1941 the Banco de México created the Office of Industrial Investigations whose mission was to research the country's renewable and nonrenewable natural resources and the industries related to these fields and other industries, as well to promote the Mexican economy. In this same area the Ministry of Economy in 1948 created the Laboratorios Nacionales de Fomento Industrial (National Laboratories for Industrial Development).[54] And in 1950 the Banco de México itself created the Instituto Nacional de Investigaciones Tecnológicas (National Institute for Technological Research) in collaboration with the Armour Research Foundation to carry out research on natural resources with the potential to be used by industry, such as food, fiber, solid fuels, and chemical products, and to provide training to graduate students.

Other changes starting as far back as September 1, 1939, were due to the war in Europe, and consequently a new national and international scenario for Mexico. In particular, the war significantly changed diplomatic and economic relations with the United States that produced direct economic benefits and new terms in Mexico's relationship with its northern neighbor. The issue of continental defense had already been discussed in meetings promoted by the United States, in which Mexico participated, held in 1938 in Lima and Havana in 1940. When the Japanese attacked Pearl Harbor on December 7, 1941, the United States became a belligerent nation; Mexico would as well in May 1942 after German submarines sunk two Mexican oil tankers. By then a series of negotiations was already taking place between the two countries that led to agreements to renegotiate Mexico's debt and on other issues such as oil expropriation, in which there were serious disputes, thereby creating a new relationship between the two countries. This produced relief in the pressure exerted by the Mexican debt on its national finances, and since the US economy had already begun the process of converting its industry to the production of weapons, it became clear Mexico would have to provide strategic raw materials and manufactured products. The country's accelerated momentum toward industrialization and economic growth during the ensuing decades dates from that time, and by 1952 its physical volume in the transformed industries was two and a half times greater than it had been in 1930.[55] Exports to the United States allowed Mexicans to obtain foreign currency for the acquisition of industrial machinery and equipment. The monetary and credit expansion of the Mexican government allowed for increased public spending on infrastructure projects that further grew the private economy.[56] Both factors allowed unprecedented conditions to exist for industrial takeoff and for the Mexican economy's modernization and growth in general.

The case of the chemical industry is illustrative. This industry grew with the oil expropriation of 1938 because of the processes required for refining that involved chemists, many of whom had graduated from the National School of Chemical Sciences at UNAM. The training of chemical engineers began in Mexico in 1925 with a basics of chemical engineering course taught by Estanislao Ramírez, but its maturity did not begin until 1935 when Fernando Orozco introduced to the curriculum modifications he had learned in Germany to strengthen the training. Elsewhere the National Polytechnic Institute did not create its programs in chemical engineering and extractive industries until 1948.

In the first year of the Ávila Camacho government, UNAM created the Institute of Chemistry with the support of El Colegio de México. It later gained support from US institutions: the Rockefeller Foundation, the Promotion and Coordinating Commission for Scientific Research, and the Inter-American Committee for Scientific Publication in New York.[57] Several industries emerged from the research done by the Institute of Chemistry on salt lakes, including the Sosa Texcoco industrial plant created in 1943 to produce caustic soda, carbonate, and sodium bicarbonate from the salts at the bottom of the lake.[58] As a result of the expropriation of German-owned companies, the National Chemical Pharmacy, a government company, was established. In 1944 Syntex was founded, which achieved the synthesis of progesterone with Mexican diosgenin. In 1947 it linked with the Institute of Chemistry to develop a business project that by 1950 made Syntex the world leader in producing steroidal hormones, highlighting the role played by the Mexican researcher Luis E. Miramontes, whose work led to the production of the first oral contraceptive.[59] Starting in 1945 the institute began publishing the *Boletín del Instituto de Química* to publicize the work of its researchers on natural products and other Mexican chemicals, boasting that it was written solely in Spanish. Starting in 1950 the chemical industry was able to start producing in Mexico basic chemical products, in addition to intermediate products.

The electrical industry, another example, had a remarkable development phase during the war years because of the impossibility of acquiring from abroad equipment and spare parts necessary to keep power plants and systems functioning. Engineer Alejo Peralta founded the company Electrocerámica in 1939, which later changed its name to Industrias Unidas. The company was originally designed for the production and assembly of bakelite lamp holders. In 1943 with its own technology the firm initiated the manufacture of ceramic insulators for low, medium, and high voltage. It also produced porcelain lamp holders, knife switches, nozzle insulators, circuit breakers, and lightning arresters. In 1941 the Electricity Manufacturing Company was established in the city of Irapuato and with its own technology built distribution transformers and magnet wire. Electrical laboratories also became important to the development of the industry. The laboratories of the Mexican Light and Power Company, Federal Electricity Commission, the National Laboratory for Industrial Development, and the Mexican Institute of Technological Research all provided important services to industrial engineering and development for electrical manufactures. Finally, in May 1945 the National Bank of Mexico collaborated with Westinghouse Electric Corporation and the US investment firm Kuhn, Loeb & Company to create Industria Eléctrica de México, which at the time was the largest and most modern industrial electric company in Mexico. By 1947, with the help of a large number of Mexican mechanical engineers and electricians trained in the United States, this company was already producing motors, transformers, generators, radios, refrigerators, and other household electrical appliances. The latter had a great impact on much of the Mexico's way of life.

The Monterrey Iron and Steel Company, founded in the early twentieth century, began expanding in 1941 and 1942, constructing a second blast furnace. In that same year the company Altos Hornos de México was born with government and private funding, the latter coming from national and foreign investors. Its equipment was acquired in the United States, some of which was highly restricted due to the war. In exchange the company committed to the Washington Maritime Commission to sell thousands of tons of steel plates to manufactures building US ships for the war effort.[60]

Another vital ingredient of industrial production, sulfur, became important during this time. Sulfuric acid had become required in a variety of industrial processes and the production of synthetic fertilizers. In 1946 the US-Mexican jointly invested Azufrera Panamericana company started to exploit the sulfur domes in the Isthmus of Tehuantepec. A 1948 authorization to expand a mining area that had been granted in 1943 to former revolutionary official Alfredo Breceda and Manuel Urquidi reaffirmed what was already a well-established industrial development policy by that time: the industrial development and economic future of the country depended on utilizing the raw materials it possessed, which is why political leaders and many technical experts pushed to build the sulfur extraction industry on a massive scale.[61]

At the same time, with increasing number of hectares irrigated by hydraulic works, it became necessary to maintain soil fertility to promote intensive high-yield crops. This meant putting in practices to better sustain the use of water resources and soil quality, especially in arid and semiarid regions. Therefore in 1942 the Soil Conservation Department was created in the Directorate of Agrology of the National Irrigation Commission to study and propose measures to prevent erosion.[62] By 1946, as expressed by the commission's executive member Aldofo Oribe Alva, it was already an institutional policy to create "scientifically prepared programs annually to take into account a crop rotation system to maintain the fertility of soils, the use of fertilizers, etc." Likewise, the department acted to industrialize agricultural products in the same irrigation districts, using the electricity produced in them.[63] For the production of chemical fertilizers and the improvement of the chemical composition of the soils, in 1948 the parastatal Guanos y Fertilizantes was given authority to synthetically manufacture chemical fertilizers, thereby producing a "technological leap in the superphosphate manufacturing process and the synthesis of ammonia."[64]

Another area where attention was paid to the country's technological and economic development after 1940 was education and university research. In 1943 the renowned physicist Manuel Sandoval Vallarta became general director of the Insituto Politécnico Nacional. He developed new regulations and helped establish a movement toward industrial research and new technology-centered careers. This direction became codified in the IPN's first Organic Law in 1950. Technological institutes were also created in Durango, Chihuahua, Jalisco, Coahuila, and later expanded to other regional centers. Thus the relationship between the state and the university, as well as with other institutions, was augmented and normalized since the educational and training effort of technical and scientific professionals was necessary for the objectives of the national economic policy, and was demanded by the country's participation in the war effort.

Moreover, during the war numerous scientific, technological, and medical institutions of importance were created out of the country's industrialization and social development: the National Committee to Fight Tuberculosis (1941), the UNAM Institute of Mathematics (1942), the Tonanzintla Astrophysical Observatory (1942), the Institute of Medical and Biological Studies of UNAM (1942), and the Commission for the Promotion and Coordination of Scientific Investigation (1942). In 1943 the Children's Hospital of Mexico was opened and the Instituto Tecnológico y Estudios Superiores de Monterrey and Universidad Iberoamericana, both private institutions, appeared. In 1944 the Mexican Institute of Social Security was created, as was the National Institute of Cardiology and the Syntex company laboratories. The Hospital for Nutritional Diseases came into existence in 1945. These institutions dramatically expanded technological and scientific education and, in turn, the growth of these sectors in Mexican society as a whole. This change increased the rationalization of Mexican society and its connections to global economies, professions, and political circles.

The development of science, technology, and technocracy in Mexico has been influenced by politics stretching back to the colonial period, but it drastically changed during the revolutionary and postrevolutionary eras. As in the Porfirian period, foreign partnerships were encouraged and appreciated; nationalism and exchanges with foreign powers were not seen as mutually exclusive. However, there were significant differences between the Porfirian and subsequent periods that resulted from policies directly related to the revolution, such as agrarian reform and the greater incorporation of urban workers into the state. The state played an even more significant role in guiding this vision of a technical and scientific future. This trend was solidified in six-year plans in which industry and government partnered with scientists and engineers to promote progress under a nationalist mantle.

Not all changes were internally driven. Many technological and political developments originated in foreign lands but drew Mexico into them. The six-year plans themselves were influenced by similar state plans in other countries, including the Soviet Union. World War II provides the clearest example of global affairs impacting Mexican development. Not only did Mexico's entry into the war and its strategic alliance with the United States increase the production and export of certain resources, but it also spurred manufacturing alongside scientific and technological exchanges. These transactions dramatically affected Mexico's economy, but also its culture and educational systems.

There is plenty to critique about the results of these developments, as many contributors to this volume do. But the significance of the changes in science, technological development, and technocratic politics is hard to overstate. They were in many ways drastic, a massive transformation of much of Mexican society within a matter of decades. These changes expanded Mexico's middle class, laid the foundations of Mexico's current economy, and provided increased access to education, goods, and services.

CHAPTER 6

TECHNOCRATIC DIPLOMACY

Constitutionalist Engineers as Diplomats to the United States

J. JUSTIN CASTRO

With the Mexican Revolution ongoing and US military forces building roads and chasing Pancho Villa in Chihuahua, Alberto Pani, Luis Cabrera, and Ignacio Bonillas together addressed the Academy of Political and Social Science and the Pennsylvania Arbitration and Peace Society in Philadelphia. They were in the United States serving as diplomats for Villa's rivals, the Constitutionalist forces led by Venustiano Carranza. Specifically, they made up the Mexican half of a formal bilateral committee to maintain peace between the United States and Mexico. An unexpected battle between US and Constitutionalist forces in the small town of Carrizal, Chihuahua, on July 21, 1916, had brought Mexico and the United States to the brink of full-out war. But these diplomats had come to Philadelphia to discuss more than their mission to maintain peace. They came as fellow scientific thinkers to share "unbiased facts." They gave presentations on hygiene, infrastructure, and municipal governance. To the applause of attendees, the president of the academy, Leo S. Rowe, a respected political scientist and Latin Americanist, wished Pani, Cabrera, and Bonillas success in building a "Mexico prosperous, progressive, independent, and sovereign."[1] Using their technical skills, appreciation of science, and progressive values, Carranza's agents obtained the backing of many of their US counterparts, who helped secure further political and popular support for the Constitutionalist cause.

As *Technocratic Visions* exemplifies, historians of Mexico have recently given significant attention to the rise of technocrats and scientific management. During the Porfirian and revolutionary eras, educated elites and government officials increasingly strove to make Mexican society more legible, creating libraries of statistics, city plans, maps, blueprints, and "scientific" policies.[2] This emphasis on scientific management and progress was, of course, not unique to Mexico, but the result of the spread of intellectual and technological trends radiating since the Enlightenment from empires and industrializing powers in the North Atlantic.[3] Mexican political leaders strove to paint Mexico as a partner in material and moral prosperity, and intellectuals from the industrial North were often eager to spread their models of rational, capitalist progress—to build a world in their image.

Little of the historical literature on technocrats, outside of some pieces about the Cold War, has focused on technocratic consular agents or diplomats. In their book about expertise and technology in Latin America during the Cold War, historians Andra Chastain and Timothy Lorek discuss networks of experts buzzing "beneath the conflicts waged by diplomats and militaries."[4] Yet as technologies became more sophisticated and abundant in twentieth-century societies, engineers and other scientific experts were increasingly involved in diplomatic and military affairs themselves.[5] This trend accelerated during the Cold War, but it began well before then. As Lucero Morelos Rodríguez and Francisco Omar Escamilla González briefly discuss in Chapter 2, engineers Gilberto Crespo y Martínez and Francisco Díaz Covarrubias served the Mexican government of Porfirio Díaz as diplomats. In those roles they were important to the transfer of information about new technological trends and as embodied representatives of Mexican modernity. As agents of industrial enterprises built steam and plowed their way across more parts of the world, the interaction between states intensified within a web of new communication and transportation networks. Diplomatic affairs became more complicated and included more issues of a technical nature. So it's not surprising that governments began to incorporate technocrats into their diplomatic missions.[6]

The focus of this essay is on technocratic diplomats who served the Constitutionalist cause during the Mexican Revolution. It builds on my previous work on Modesto C. Rolland and other revolutionary engineers to emphasize that this particular group demonstrates an important development: an increase in the use of Mexican technocratic diplomats who played on US rationalist tendencies to obtain political and military support.[7]

Although scholars have not paid much attention to Mexican technocratic diplomats, they have looked at US experts in diplomatic roles in Latin America. The works of historians Helen Delpar and Ricardo Salvatore discuss US academics, including Rowe, who served in informal diplomatic roles in South America during the first half of the twentieth century.[8] Salvatore argues that these US experts engaged in "knowledge enterprises" and that their activities should be "considered ancillary activities in the making of imperial hemispheric hegemony."[9] Chastain and Lorek agree.[10] Information-gathering academics and US-Latin American foreign policy went hand in hand. The rise of Latin American studies programs in the United States was intertwined with US cultural empire. American officials measured Latin American societies against those in the United States, and in the process they purposefully (as well as unintentionally) promoted the "U.S. way of life as a model to be imitated or replicated by South American societies."[11] What is equally true is that many Latin American officials were already interested in positivist and progressive philosophies originating in Europe and elsewhere about development before US scholars flooded the lands south of the Rio Grande. Many Latin American officials also understood US proclivities and used US interests to their own advantage.

In an examination of the consular and press operations of influential Constitutionalist engineers who served diplomatic roles in the United States, I argue that these technocrats learned to use the progressive and scientific predispositions of the Woodrow Wilson administration and US academics, journalists, and activists to their advantage.[12] Mexican technocratic diplomats convinced many prominent American progressives that Constitutionalist leaders were the Mexicans most like them and, in turn, that the Constitutionalists were the faction most worthy of US support. This effort was especially strong in the wake of Constitutionalist military victories over the forces of Pancho Villa throughout 1915. Certain Mexican engineers, including Pani and Bonillas, proved adept at using their technical know-how, progressive bonafides, and self-proclaimed apoliticism to sell Constitutionalist propaganda as unbiased truth to news outlets and to US progressives who helped spread their information. In other words, Constitutionalist technocrats serving as diplomats played on US progressive proclivities, convincing them to provide support and to spread Constitutionalist propaganda.

THE EDUCATION OF A REVOLUTIONARY ENGINEER

As Jayson Porter points out in Chapter 7, engineers had become important members of the intelligentsia within the different factions competing for power in the Mexican Revolution. Rarely soldiers, these engineers constructed networks, plotted the eventual reconstruction of the Mexican nation-state, and served as press agents and diplomats. All factions incorporated engineers and other technocrats, but Carranza's Constitutionalists possessed more of them. While combatants fought and died in Mexico, consular and commercial agents waged a war of guile and propaganda, attempting to persuade influential people within foreign nations to throw their support behind one faction or another. Because of its weapons, large economy, and proximity to Mexico, the United States was where these diplomats fought most contentiously. The Constitutionalists' edge in technocratic expertise ultimately proved important to their successful drive to obtain recognition and support from the United States. It provided a sense of rationality that, despite Constitutionalist nationalist rhetoric, was compatible with US academic communities' and capitalists' desire for an orderly economic recovery.

This resemblance between Constitutionalist foreign agents and US progressives was not originally a ploy, though it did become one. US progressives found many of their Constitutionalist counterparts similar because they shared comparable and sometimes overlapping educational and professional networks. Almost all the Mexican technocrats discussed in this chapter were educated during the Porfirian era, some of them in Mexico, some in the United States or France. They were born between 1876 and 1881, the one exception being Bonillas, who was born in 1858. Cabrera was a baker's son who obtained a law degree in Mexico City from the National School of Jurisprudence in 1901 and quickly became a successful lawyer, an important voice on agrarian reform, and a critic of Díaz. Pani trained at the Aguascalientes Scientific and Literary Institute and then the National School of Engineering where he became a star pupil in civil engineering, working on potable water projects. Bonillas was born in Sonora but mostly raised in Tucson, Arizona. He obtained his engineering degree at MIT and married into a prominent Arizona family before exchanging a career as a mining engineer to become mayor of Madelena, Sonora, and an active member of the Mexican Geological Society. Other important Constitutionalist engineers who served diplomatic roles in the United States had similar backgrounds: Modesto C. Rolland (Escuela Nacional de Ingeniería), Francisco Urquidi (École Centrale des Arts et Manufactures), and Juan Francisco Urquidi (MIT).[13]

Engineers trained in the 1890s and early 1900s in Mexico City were greatly influenced by their mentors. These mentors, including some of the same engineers from the late 1800s discussed in the chapters by Morelos and Escamilla, Garza, and Vitz, had been strong proponents of liberalism, scientific management, and progress. They shared a number of ideas put forth by Mexican positivists like physician Gabino Barreda, who wanted to introduce a "new system of scientific preparatory education" and who promoted a "mental emancipation" from practices connected to the Catholic Church, communal Indigenous communities, and government monopolies. Similar to their Porfirian predecessors, revolution-era engineers argued that political and social changes were not separate from scientific practices; they were intertwined.[14] Both groups contended that political order and technological development were key to social progress. They saw themselves as a crucial component of that progress because they designed physical and social spaces. They created built environments that would improve the health and productivity of Mexico's citizenry, and they plotted the infrastructure that drew together disparate parts of the nation and the nation to the world. To carry out this vision, they built on experts and writings from the United States, but also from a century of works produced domestically and taken from Europe.

One way in which engineers such as Pani and Rolland, who emerged in the first years of the 1900s from the Escuela Nacional de Ingeniería (National School of Engineering), differed from their predecessors was their more vocal promotion of electoral democracy. Mexico's lack of democratic governance had not been lost on their Porfirian predecessors. They, along with other "scientific" planners, including the polemic engineer Francisco Bulnes—who supported Díaz's sixth reelection run in 1903 but lambasted Mexicans who promoted the president-turned-dictator's candidacy as a symbol of democracy—had no reservations about saying that the democratic ideals in the liberal Constitution of 1857 had not come to fruition. Engineer Ricardo García Granados wrote in 1906 that the crafters of the 1857 Constitution had failed to understand the Mexican people's inability to successfully take up democratic practices. García Granados and Bulnes represented an intellectual current that praised liberal secularization and the material progress of the Porfirian era but argued that a democracy could not truly exist in Mexico until its population was literate, educated, and unified enough to make a genuine democracy work.[15] The widespread participation of young engineers in the anti-reelectionist clubs who supported Francisco I. Madero's run for the presidency in 1909, and who participated in the revolution that followed, shows that they believed more than their mentors that Mexico was ready, or could quickly be made ready, for electoral democracy.[16] Their outspoken promotion of democratic politics—despite their often elitist and instrumentalist perspectives—complicates claims that technocrats have been undemocratic.[17]

Revolutionary engineers sometimes carried a sense of arrogance, like many of their predecessors; yet alongside their promotion of electoral democracy, many of them espoused rhetoric circling around the globe about social justice, the alleviation of urban plight and poverty, and, to some extent, support for women's rights. There was more of a socialist bent to their brand of liberal capitalism, even if a close reading of their works shows abundant racism, classism, and patriarchal paternalism (similar to their US counterparts). This marriage with global progressivism was particularly abundant in those engineers who had aligned themselves to Madero during the revolution and then left Mexico for exile in the United States after his assassination in 1913. Many of the engineers who rose to prominence during this period were also vocally nationalist, promoting the "Mexicanization" of Mexico's infrastructure and the hiring of Mexican engineers. The Engineer's Club, associated with the National School of Engineering and founded in 1908, promoted the Mexicanization of Mexico's National Railways and communications.[18] But in their promotion of Mexican sovereignty they regularly appealed to the incorporation of foreign tools, education models, and governmental policies prescribed by Western technocrats.

That so many of these engineers served the Constitutionalists should not be surprising. As supporters of Madero, they had become enemies of the Victoriano Huerta regime that had Madero assassinated. When Villa and Zapata split from the Constitutionalists to help form the Conventionalists, most engineers preferred the more orderly, urbane, and middle-class leadership of the Constitutionalists to the cowboy charisma of Villa and the rural and seemingly antimodern peasant armies loyal to Zapata.[19]

THE ORIGINAL REVOLUTIONARY ENGINEER-DIPLOMATS

The most prominent engineers serving diplomatic functions in the United States during the early phases of the Mexican Revolution were Francisco Urquidi, Juan Francisco Urquidi, and Modesto Rolland. All of these engineers—and all the others discussed in this essay—supported the original Madero uprising and his subsequent government. Francisco Urquidi was the only one to serve during the Madero administration in the United States. However, Juan Francisco Urquidi and Rolland joined Francisco Urquidi shortly after the assassination of Madero in February 1913. The Urquidi brothers were among the most conservative and privileged of the revolutionary engineers who took on diplomatic roles. Carranza relied on them because of their proficiency in English, their foreign education and experience, and their ability to use their pedigrees and educational networks to impress US officials and business leaders. But unlike Rolland, Pani, and Bonillas, whom I discuss in greater detail later in the chapter, Francisco Urquidi and Juan Francisco Urquidi rarely used their engineering skills as diplomatic tools. They nonetheless established a precedent for using engineers as diplomats because of the benefits they provided by their foreign training, language skills, and by the association of engineering with a modern identity that appealed to many US intellectuals and political leaders.[20]

The youngest Urquidi, Juan Francisco, was a relatively effective diplomat. His success owed more to his strong familiarity with the United States and his education than his ability to sell his technical skills or Constitutionalist reconstruction. At the turn of the twentieth century he attended high school at the Dean Academy in Franklin, Massachusetts, and subsequently obtained his engineering degree at MIT. Juan Francisco loved the United States, writing praises to his family about the vibrancy, beauty, and bustle of New England. He believed that engineering programs in the United States had transcended all others in quality because to him US education and culture were more practical than in Mexico or, for that matter, in France, where his older brother Francisco had obtained his engineering degree.[21]

After briefly returning to Mexico to work with his brothers, Juan Francisco became a leading confidential agent for Carranza in Washington, DC, following Huerta's rise to power. Urquidi spent much of his time sending and receiving news via wire services, attempting to persuade Americans to support the Constitutionalist cause, and working to obtain goods and arms. Although his engineering background surely proved useful in obtaining supplies, most coverage of him in the US press surrounded his service as a Carranza representative during the Niagara Falls Peace Conference, or ABC Conference, a meeting of officials from the governments of the United States, Canada, Argentina, Brazil, Chile, and Huerta's Mexico to discuss a possible transition of power in Mexico and to address the growing tension between the United States and Mexico. The Constitutionalists had originally refused to participate and then only did so informally, but Urquidi worked alongside other prominent Constitutionalists to relay their position, which was ultimately that they would not compromise with Huerta's government, that US intervention should end, and that they welcomed peace with foreigners who respected Mexico's sovereignty. Coinciding with the US occupation of Veracruz, it was a difficult diplomatic task.[22]

Press agents generally described Urquidi as capable, but his engineering background was not a point of discussion. After reliably relaying messages and navigating the tense diplomacy over the transition of power from Huerta to Carranza, Urquidi increasingly questioned the path of the revolution as the forces loyal to Carranza and those of Villa and Zapata split. Urquidi subsequently retired from the revolution to pursue interests in literature and literary criticism.

Francisco Urquidi, the oldest of the Urquidi brothers, had begun his engineering studies in 1895 in Paris at the École Centrale des Arts et Manufactures (Central School of Arts and Manufactures), a school after which institutions throughout the industrializing world had modeled themselves. There he obtained a degree in industrial engineering and then remained in France to take additional courses in electrical and metallurgical engineering. After facing difficulties retaining the Mexican government funding that had allowed him to stay in France, he returned in 1905 to Mexico, where he taught physics and mathematics at the National School of Engineering and did electrical work for the Ministry of Public Education. He also tinkered with inventions, mostly associated with electronic player pianos.[23]

From the onset of the Constitutionalist uprising, Francisco Urquidi worked for Carranza as a commercial agent in New York. Before the Carranza–Villa split, Urquidi focused much of his time on acquiring materials for Carranza, downplaying friction between the two leaders, and fretting about security for his family and himself. Throughout 1914 Urquidi worked to obtain goods for the Constitutionalist forces. His technical expertise was useful in this regard, though he failed in his attempt to obtain airplanes for the Constitutionalists.[24]

Urquidi occasionally wrote press pieces in which he defended the Constitutionalist cause. He was an elegant writer, and his erudition caught the attention of those with whom he interacted. An article he wrote for the *New York World* and *St. Louis Post-Dispatch*, "Let the United States Keep Hands Off Mexico and We'll Crush Huerta Soon" is exemplary. In it he argues against further US intervention while laying out reasons for the revolution, comparing it to the French Revolution. He also provides his take on some of the Mexican Revolution's leaders. The Constitutionalists would topple Huerta on their own, he declared. Urquidi lamented the old Porfirian "merchants and their wives, fattened parasites of plutocracy." While incorporating racist tropes, he sympathized with oppressed Indigenous peoples: "An unwritten law decrees that only the well-dressed may venture there [the Paseo de Reforma]. The procession rolls by, of firm assurance, never dreaming that the pavement is but a flimsy crush of usurped and fleeting power, beneath which smolders a fiery lake of popular wrath." The article was one of his few attempts to demonstrate his flair for prose but also his self-prescribed apolitical character and his scientific rationality. Describing the revolution in medical terms, he allowed that "humanity has a right to ask what the surgeon's plan is, whether your operation really offers a fair chance of success."[25] Although his attack on the privileged was a bit ironic since he came from a wealthy family and was educated in France on a stipend from the Porfirian government, he hoped his educated references to the French Revolution and his medical terminology would resonate with educated US readers interested in social justice.

US journalists were more apt than Urquidi to bring up his technocratic expertise. Alongside the aforementioned article, a reporter discussed Urquidi's excellent education in engineering, his early and successful career as an electrical engineer, his time as a professor at Mexico's National School of Mines and National School of Engineering, and his time as a Mexico City alderman when he worked on the "most important" rapid transit, sewers, and lighting commissions. But despite his talent, the journalist continued, Urquidi "took no active part in politics."[26] Even as Urquidi pled with Americans to let the Constitutionalists topple Huerta on their own, he still sought no political office. It was a line that Mexican technocratic diplomats used with regularity: they were scientific designers of a modern and more just society; they could be trusted like their counterparts in other modern governments because of their enlightened apoliticism.

However, it was not a line that Urquidi himself often used. In the lead-up to the split between Carranza and the Villa and Zapata camps, Carranza called Urquidi temporarily back to Mexico. The "First Chief" had doubts about Urquidi's loyalties. During that brief period, Urquidi's fellow engineer-diplomat Modesto Rolland ramped up a press operation in Urquidi's office that touted Constitutionalist reconstruction policies. When Urquidi returned he attempted to stop Rolland's work, writing to Carranza that it was improper and "inconvenient."[27] Carranza concluded otherwise.

Perhaps due to their mutual connection to the state of Chihuahua and shared frustration with Carranza, Urquidi and Villa joined forces following the Aguascalientes Convention. Urquidi became Villa's consul general in New York but was rarely effective diplomatically.[28] During his limited press engagements from late 1914 through 1916, he spent most of his time defending Villa's character, attempting to counter Constitutionalist press attacks, and downplaying Villa's battlefield failures even as they became brutally apparent.[29] Urquidi blamed the Wilson administration's turn against Villa on "the strong press campaign carried on in this country and in Mexico by Carranza forces."[30] He was not completely wrong. Alongside the Constitutionalists' successful military campaign against Villa, Carranza and his agents in the United States ratcheted up their media operations, in part led by Rolland, who not only ruthlessly and relentlessly criticized Villa but also consistently discussed his own work as an engineer and the reconstructionist policies carried out by technical experts in Constitutionalist territories. While Urquidi was denying Villa's defeats, Constitutionalist spokesman Roberto V. Pesqueira was discussing a "new era" for Mexico and the "opening of its railroads, mines, and factories, the resumption of legal business . . . the beginning of an epoch of prosperity for the natives as well as for the foreigners. It means the greatest industrial awakening in the history of Latin America."[31] Urquidi never worked as an actual engineer for Villa, and he never touted Villa's ideas for reconstruction. Villa dismissed Urquidi in 1916, and Francisco subsequently died of the Spanish flu in 1918.[32]

CONSTITUTIONALIST DIPLOMATS OF THE MEXICAN-AMERICAN JOINT COMMISSION

A lot had changed by the time Pani, Bonillas, and Cabrera gave their late summer 1916 talk in Philadelphia. Villa had transitioned from a public hero supported by much of the Wilson administration to a villain in the eyes of most Americans. Aside from Francisco Urquidi, most revolutionary engineers sided with Carranza. The Wilson administration tentatively sided with the Constitutionalists after they handed Villa serious military defeats, which was one reason why Villa had decided to raid Columbus, New Mexico, and spark US intervention in the first place. But that did not mean that the Wilson administration or powerful US business operators were comfortable with Carranza or that continued support for the Constitutionalists was guaranteed. When the Wilson administration sent the Punitive Expedition after Villa, the situation put the Carranza administration in a tough spot. Its leaders wanted Villa dead or captured, but they did not want to look like they were collaborating with a US invasion. Carranza again demanded respect for Mexican sovereignty and that the US military limit its expedition to a small portion of northern Chihuahua. The clash at Carrizal resulted from US soldiers challenging Carranza's demands, and it renewed tensions among Carranza, Wilson, and the US public. The Constitutionalists further ratcheted up their media and diplomatic operations.

Carranza's ardent nationalism never played well with many US business and political leaders who had grown to expect a Mexico that showered Americans with favoritism and that limited regulations on their schemes. The Wilson administration had attempted on previous occasions to find a compromise president, someone who would remove Villa and Carranza from any potential presidency and who would better appease US concerns over Villa's lack of perceived civility and Carranza's nationalism, which, even if justified, seemed to many Americans as anti-American. There were still plenty of voices calling for the Wilson administration to turn away from Carranza. Wilson was also facing an upcoming reelection.

The Mexican-American Joint Commission set out with prominent representatives and a tough set of goals. Secretary of the Interior Franklin K. Lane, former senator and judge George Gray, and John R. Mott, an evangelical theologian and secretary of the international component of the YMCA, represented the United States. The *Washington Post* labeled the Constitutionalist representatives—Cabrera, Bonillas, and Pani—as "technical men."[33] None of them was a consular or commercial agent, though all of them had served in various diplomatic missions. Bonillas would become the official Mexican ambassador to the United States in 1917. Despite regular condemnation in the United States of Mexican leaders in general, *Washington Post* journalists respected the Mexican representatives' pedigrees, making particular note of Bonilla's MIT engineering degree. The Constitutionalists were mostly interested in avoiding the expansion of US forces in Mexico and the withdrawal of the Punitive Expedition. The Wilson administration wanted to focus on the reduction of violence along the US–Mexican border and the rights of US citizens and businesses in Mexico. Neither side budged much beyond agreeing to take steps to expand border protections. Nonetheless, tempers over Carrizal cooled, and Wilson

Fig. 6.1. The US-American Joint Committee, 1916. *Source*: Courtesy of the Library of Congress. https://www.loc.gov/pictures/item/2014702648/. Creator: Bain News Service.

continued to tepidly support Carranza's forces as valid revolutionaries who represented Mexico's best attempt at a "new reconstruction."[34]

Despite the lack of agreement in the Mexican-American Joint Committee, the Constitutionalist representatives possessed something that built favor among some US progressives, something the Constitutionalists could expand further public relations campaigns on: their faction possessed the greatest number of educators and technical experts. Their civilian leadership represented modern middle-class values better than any other group vying for power in Mexico. They ardently defended Mexico's sovereignty and were not always favorable to US international businesses, but they, better than their alternatives, possessed skilled experts who could potentially reestablish a functional government and infrastructure that could modernize Mexico and subsequently benefit the United States.

Amplified by Constitutionalist propaganda bureaus in the United States, many US progressives promoted the professionalism of Carranza's agents. These technocrats were men who could, in the words of the widely read novelist and activist Mary Hunter Austin, allow Mexicans to "reestablish themselves in the scheme of social evolution." Austin called Pani "a very pretty man whose bright, reflective eyes hint without revealing his extraordinary acquaintance with the physical phases of every plan of national betterment."[35] He was an engineer and manager of the Constitutionalist-controlled railways with the "broad ability of the best type of modern businessmen."[36] Cabrera had been serving as minister of finance and had in large part drafted the Constitutionalist Agrarian Decree of January 1915. Bonillas was serving as Carranza's minister of communications; he was an MIT graduate who had married a US citizen. He possessed "initiative, executive ability, and political experience in addition to his technical qualifications."[37] To continue with Austin: "It is significant to note that the three men of whom have been entrusted the conduct of the Mexican interest are all very much of the type we would have chosen ourselves."[38] These Constitutionalist representatives reminded certain members of America's middle and professional classes of themselves.[39]

Pani, Cabrera, and Bonillas in many ways had more success in obtaining support outside of the Mexican-American Joint Commission than in it. Like several other Constitutionalist agents, they increasingly engaged in speaking events with US organizations. When they talked to the American Academy of Political and Social Science and the Pennsylvania Arbitration and Peace Society, they played up their scientific and technical capabilities. Pani focused his discussion on hygiene and city planning. After all, he had recently served as Madero's undersecretary in the Ministry of Public Instruction and Fine Arts and as a public health official for the Federal District Department of Public Works. For Pani, a clean society was a civilized society. It was filth, disease, and death that bedeviled Mexico, keeping its people from entering the ranks of wealthy and advanced countries.[40] In 1916 he had gathered his thoughts into a tome he titled *La higiene en México*. He would soon thereafter publish an English-language translation in the United States and England.[41]

The book talked a language familiar to progressives around the world. Comparing mortality rates from a wide range of cities, mostly in the industrialized West, but also in Cairo and Madras (today's Chennai), Pani concluded that Mexico City had been for several years "the most unhealthful city of the whole world."[42] True or not, Mexico City had significant health problems: pollution, poor water quality, flooding, low food safety standards, and poor housing conditions. He blamed these problems on a myriad of natural and human-made causes. He even bemoaned military officials who dominated much of the revolutionary bureaucracy and often privileged their own personal gain over the well-being of the millions of workers and children whom revolutionary leaders had promised to uplift. Pani recommended an increase in sanitary administrators, laws for compulsory sanitation in cities, and "elevating poorer classes morally, economically, and intellectually."[43] US progressives nodded approvingly of Pani's call for increased legislation and moral uplift.

What was just as important to Pani's influence in the United States as his solutions for improving Mexico's hygiene was the language he used:

Mexico City combated "primitive" cleaning operations; filth was a "heinous sin"; "the tenements and lodging houses of Mexico, homes of the great majority of the metropolitan population" were sites of "physical and moral infection." Pani stressed the need for modern sanitary codes, improved municipal organization, and the "beneficial influence of sunlight on the healthfulness of dwellings." In Pani's envisioning, "the house of the future, socialized, a living thing, sweet, educating, and consoling, shall be the true nest of human couples who may wish therein to improve the species, and launch it triumphantly forth down the ringing grooves of time." He focused on the urban environment, though he hoped that urban studies would eventually help rural communities as well. These were words and sentiments shared by British and US reformers and cast in their language.[44]

So, it is not surprising that Pani's obsession with hygiene and progressive jargon found a welcome audience among many US intellectuals. The ideas that influenced Pani, after all, had mostly come from European and US scholars and activists. As the revolution evolved (or devolved) into civil war, Pani's credentials as an engineer, public official, and sanitation scholar served him well during his interactions with US academics and public officials. In his Philadelphia speech Pani laid out the same arguments he made in the book, which he equated to why the Constitutionalists, especially people like him, were the right people to lead Mexico. Lamenting the violence and destruction of the revolution, even if justified to some extent, Pani argued that rebuilding a better Mexico required a core of individuals capable of improving living conditions and of establishing a well-functioning government. And the Constitutionalists were led by people like Pani: educated men, technocrats with middle-class and progressive tendencies, people who built infrastructure and studied hygiene. They were civilized people. Their enemies, including the forces loyal to Villa and Zapata, were, according to Pani, less orderly and more destructive, even reactionary. They were antimodern. Reconstruction would take grand ideals, something all revolutionary factions declared, but it would also take technical expertise, which is something Constitutionalist agents advertised with zeal.

Cabrera was not an engineer, but he was an architect of Constitutionalist policies and sometimes couched his ideas in technocratic terminology. In his speech he laid out his condemnation of the earlier Díaz administration and presented new ideas for reconstruction. He too was particularly keen to take advantage of this academic audience. He was glad to have the opportunity to make himself "heard before a scientific and scholarly public, free from prejudice and interested in the Mexican situation."[45] Cabrera let the listeners know that unlike biased newspapers and partisan Mexican ex-pats, he offered a "purely scientific" narrative that applied "sociological criteria." During his time in the United States, Cabrera spoke at a number of other, similar events, working to influence progressives by highlighting the material progress underway in Constitutionalist-controlled regions of Mexico and by stressing his objectivity even though he was a cabinet member in Carranza's government.[46]

Bonillas's speech was short. He expressed regret for the revolutionary violence and then focused on developments of which US academics and pacifists would approve. He applauded his fellow Constitutionalist planners for developing reconstructive works in the face of war and instability. Echoing President Wilson, Bonillas promoted the idea that infrastructure and peace went hand in hand. He promised his audience that progressive municipal and state reforms were underway in areas that the Constitutionalists had gained control. Carranza, Bonillas continued, was attempting to "restore order at the earliest possible moment."[47] Elections were underway, and a new constitutional convention would start in a matter of days on November 20, 1916.

Bonillas also spent time in front of journalists talking about bringing order out of revolutionary chaos through physical reconstruction. An article in the *New York Times* praised Bonillas's MIT education while expounding on his work on infrastructural development. Bonillas highlighted the Constitutionalists' increasing domination of wireless or radiotelegraph stations. The journalist poured on adulations, touting Bonillas's work reestablishing railroad, telegraph, maritime, and mail services.[48] Bonillas sent an even lengthier account to the *Mexican Review*, a monthly periodical published by Carranza's US ally and newspaperman George Weeks. In it Weeks not only praised Bonillas's work reestablishing and improving infrastructure and communications, but also Pani's operation of the railways.[49] Although Bonillas sometimes had a more difficult time swaying opinion in Mexico, his engineering abilities, in addition to his Boston-influenced accent and well-connected wife from Arizona, swayed certain Americans.

The Constitutionalists used the *Mexican Review* to tout the technical bonafides of all the members of the Mexican-American Joint Commission and other Constitutionalist agents. Luis Cabrera was the "leading exponent of fiscal reform." Bonillas was the exceptional MIT graduate-turned-secretary of communications and public works. Pani was the president of the National Railway Lines of Mexico.[50] The Carranza government had expanded its control of railways from 2,000 miles of track to 13,000 miles. They were never more "ably and honestly handled and operated" than under Pani's supervision.[51] Weeks highlighted the reconstructive efforts in Yucatán under General Salvador Alvarado and his advisers "big in brains and enterprise."[52] To offset criticism of Carranza's outspoken economic nationalism, Weeks promoted Mexico City streetcars and the reopening of oil enterprises, mines, and other industries of interest to US businesses.[53]

Military victories in Mexico aided Carranza's forces more than anything else, but winning on the battlefield was not the only way Constitutionalists influenced Americans. Technocrats serving as diplomats also gained support because they looked and acted similar to how many US progressives saw themselves: well educated, well dressed, idealistic, but also practical, and desirous of order. Modern states required technologically adept militaries that could monopolize violence, but also university-trained techno-bureaucrats who could rationally design infrastructures, economies, and social order.[54] In a healthy, capitalist order, that also meant the promotion of a strong middle class. Carranza's propaganda machine hit these points constantly. To turn back to Pani's Philadelphia address, the revolution was a convulsion, a symptom of the "loathsome corruption of the upper classes, and the inconscience and wretchedness of the lower."[55] Carranza's forces, as sold by Pani, represented a science-based and more-balanced middle road, and that is what US academics and progressives like Rowe wanted to see and hear. Or, in the words of US journalist and Constitutionalist sympathizer Timothy Turner, Carranza's diplomats in the United States "did not want a reaction like that represented by Huerta, neither did they want a proletarian dominance that would stamp out what gains Mexicans had made in culture, in the art of living, in developing a middle class."[56] Instead they offered a capitalist, albeit more nationalistic, Mexico led by civilian lawyers and technocrats.

MODESTO ROLLAND AND THE MEXICAN-AMERICAN PEACE COMMITTEE

While Pani, Cabrera, and Bonillas discussed and debated with their US counterparts, Modesto Rolland was serving in a similar though less official binational group created to deal with the same issues. US progressives associated with American Union Against Militarism approached Rolland to join with prominent US intellectuals to form a committee to establish a voice for peace in the immediate aftermath of Carrizal. They reached out to Rolland because he had become a known figure, a Constitutionalist progressive who had been studying educational, municipal, and agrarian reforms while operating a media operation.

Following the assassination of Madero in February 1913, Rolland had fled to the United States but had returned quickly to the Mexican border to meet with Carranza, who asked Rolland to return to the United States to research education systems and municipal governance in New York City. It was in this role that Rolland first sought to learn from and then influence US progressive intellectuals. For example, Rolland worked with William H. Maxwell, the first superintendent of schools of the City of New York. Maxwell, an Irish immigrant, had become influential in US education circles. He had written widely read textbooks on grammar and several articles on topics ranging from managing secondary education to dealing with students' parents. Maxwell was nearing the end of his professional career, but he was still tirelessly working, with mixed success, on improving high school conditions and attendance, especially among the city's many immigrant communities.[57] Rolland also studied schools in Wisconsin, which appear to have left the greatest impression on him. He sent reports back to Carranza suggesting that any future Constitutionalist administration should model its rural education schema on the system in Wisconsin; its "vocational nature" would best suit Mexico.[58]

Rolland's research in New York and Wisconsin had left an impression on the US progressives with whom he interacted. Several newspapers picked up on his endeavors, pointing to "the commission to study the free school systems of the United States" headed by Rolland as exhibiting the progressive nature of the Constitutionalists. Once in power, the *Rock Island Argus* argued, a Constitutionalist education system embracing a US model would prove a positive contrast to the "desolation that would have resulted had we [Americans] gone into Mexico and reduced the country to ruins."[59] Wilson supporters in the US Congress praised Rolland's investigations into schools as a sign that the Constitutionalists were genuine in wanting to improve Mexican education.[60]

During his studies of municipal governance, Rolland continued to seek out US progressives in New York. He also followed developments in Glasgow, Scotland, which many US progressives had praised as an example of how an active city government could clean up a city and make it better serve working people. Rolland took up the progressive calls for the initiative, referendum, recall, and hygienic living, subsequently publishing plans for a Constitutionalist Mexico that would incorporate progressive visions of municipal governance and education from Britain and the United States in a chapter of *Carranza and Mexico*, which was written largely by Rolland's associate Carlos di Fornaro, the New York–residing Italian journalist and cartoonist who once ran a newspaper in Mexico City condemning then-president Porfirio Díaz.[61] Matthew Vitz is probably right in his chapter in this volume that Rolland's verbal tirade about the lack of competency and graft among Mexico City council members during the 1922 water crisis fueled an antidemocratic trend supporting technical experts over local, democratically elected bodies, but Rolland spent much of his career arguing in support of free municipalities, progressive municipal reforms, and an end to clientelist politics in Mexico. He was among the leading public champions of these ideals. The year before the 1922 water crisis, Rolland had written an influential book specifically promoting these practices, and he helped author some of the first serious governmental policies on these exact issues.[62]

Supplanting Francisco Urquidi and working alongside di Fornaro in late 1915 and early 1916, Rolland organized the Mexican Bureau of Information and the Latin-American News Association, a pro-Carranza media organization that pumped out pamphlets, gave interviews, and provided materials for US journals and newspapers. The Mexican Bureau of Information regularly sent out the *Mexican Letter*, which contained brief essays condemning Villa and promoting Constitutionalist positions, to US news outlets. The Latin-American News Association published stand-alone essays written by Rolland and other Mexican and US allies of the Constitutionalists. Issues included discussions of reconstructive policies in Yucatán, the "religious question," municipal governance, labor policies, national sovereignty, and Villa's shortcomings.[63] Emphasizing his importance to Carranza's media operations in the United States, Rolland claimed that the mailing list he used to distribute information included approximately twenty-five thousand prominent people and organizations.[64]

Rolland received considerable attention from US progressives. They were impressed by his education, progressive prescriptions, and the fact that Rolland had not only drawn up policies but had carried through with them in parts of Mexico. In addition to being a propagandist, Rolland actively used his engineering skills, working as a communications officer and as the head of the local agrarian commission in Yucatán. In a full two-page article on Rolland that focused on his work and policies aiding the military governor of Yucatán, Salvador Alvarado, the *New York Times* heaped adulation on Rolland, the destroyer of preconceived notions about Mexico. According to the journalist, Rolland talked and thought like John Haynes Holmes, the famous Unitarian minister and cofounder of the National Association for the Advancement of Colored People (NAACP) and American Civil Liberties Union (ACLU). The reporter also compared Rolland to the renowned French writer Romain Rolland and the British novelist and social commentator H. G. Wells. The article highlighted the influence of progressive policies in New Zealand and Massachusetts on Rolland but went further in emphasizing Rolland's insistence that these ideas were becoming a reality through his work in Yucatán.[65] The journalist painted Rolland's ability as an engineer to personally bring plans to physical fruition in a war-torn and culturally diverse Mexico as something heroic.

Rolland's influence was on clear display during his work with the Mexican-American Peace Committee. While the Wilson administration contemplated a formal binational discussion, the American Union Against Militarism—a pacifist organization headed by Crystal Eastman, a prominent progressive and labor activist—sought out Rolland as one of three members to represent Mexico in talks with prominent US intellectuals and organizers to form a joint commission to thwart calls for war. The other invited Mexicans were Dr. Atl, a prominent artist, labor activist, and newspaper editor, and Luis Manuel Rojas, the director of the Biblioteca Nacional (National Library of Mexico). The US representatives who served on the committee included David Starr Jordan, an ichthyologist and the former president of Stanford University, Amos Pinchot, a lawyer and prominent reformist: Moorefield Storey, the first president of the NAACP, and Paul U. Kellogg, editor of the influential progressive periodical *The Survey*. Not surprisingly, the group called for peace as the best path for Mexican-US relations; it also called for labor reforms, women's rights, improved irrigation, equitable taxes, land redistribution, educational exchange programs, and greater Mexican control of the petroleum industry in Mexico. These were changes that Constitutionalist and US progressive reformers alike had previously stated as necessary for the welfare of Mexico and peace. Arguably, little of immediate substance, at least regarding policy changes, resulted from the meeting, but like its more official counterpart, it did provide a countervoice to war. More important for the Constitutionalists, the meeting helped solidify US progressive support for the Constitutionalists as the best group to reestablish order and reconstruct Mexico.

Being the vice president of the committee (Storey served as president), the only Mexican representative fluent in English, and someone who had been an active participant in agrarian and labor reform, Rolland particularly influenced his US counterparts. In *The Survey* Kellogg praised Starr and Rolland, the "lower California Engineer identified with the revolution, and for some time resident of New York," as being crucial to efforts to "stem the tide toward war."[66] Kellogg, who shared with Rolland a passion for the teachings of US political economist Henry George, became a vocal spokesman for Rolland's work and the Constitutionalists more broadly. *Forum*, another widely read US journal, published a twelve-page article by Rolland in July 1916 in which he laid out his case for Constitutionalist causes and plans for reconstruction.[67] Rolland became influential in progressive circles connected to the men and women associated with the American Union Against Militarism, which in turn helped legitimize the information he provided as something more akin to "the truth" than propaganda.

This influence stemmed from Rolland's position and English-language skills, but also because urban, privileged Americans saw Rolland as someone like them: educated, refined, capable, but also progressive and interested in social justice. Many US progressives were impressed by his technical expertise. He not only talked about reconstruction, infrastructure, and an improved Mexico, but was actually drawing the blueprints, shaping land reform policies, and working on communication networks. As many progressives saw it, if the United States was going to build a longer-lasting, more lucrative, and mutually beneficial relationship with Mexico, it would have to be with people like Rolland, people who understood how to build a more unified and modern Mexico.

It is difficult to gauge with precision how much Constitutionalist technocratic agents influenced the Wilson administration or US public opinion. Undoubtedly it was Constitutionalist military victories that most influenced the Wilson administration's decision to recognize the Constitutionalist government. Beyond a couple of influential and unscrupulous newspaper owners, Huerta's representatives and propaganda met with a cold reception in the United States. He was too obviously a usurper. Villa was more savvy at manipulating US media outlets. Some of the United States' most notorious newspaper moguls, including William Randolph Hearst and Harry Chandler (who both had extensive properties in Mexico), supported Villa. He worked with the Mutual Film Company, which paid him to provide material for their newsreels and a fictional film, *The Life of General Villa*. For a while, Villa possessed substantial influence on the US Left, including on the popular journalist John Reed and on community organizer Mary Harris "Mother" Jones.[68] But despite popular portrayals of Villa as a sort of Wild West Robin Hood, these same imaginings also portrayed him as someone unrefined, unpresidential, unmodern, and at times unscrupulous. Constitutionalists took advantage of this image, portraying Villa as ignorant, savage, and easily manipulated.[69] They portrayed themselves in contrast.

Villa's main agents in the United States never did enough to counter Constitutionalist propaganda. They spent much more time defending Villa's actions and personality than on painting a clear picture of what a Mexico in which Villa was victorious would look like. In part this was due to the divisions within the Conventionalist camp, whose leadership became murky following their ouster from Mexico City in 1915. But it also had to do with Villa's choice of representatives. The most prominent of them were foreigners with questionable reputations and loyalties: George Carothers (a corrupt US agent) and Felix Sommerfeld (a German spy).[70] According to historian Friedrich Katz, these men ultimately hurt Villa's cause: "Their loyalty was to their pocketbooks rather than to Villa, the revolution or to Mexico."[71] Like Villa's US ally Hugh C. Scott, Carothers was more conservative than Wilson and aspired to profit from his relationship with Villa while increasing and protecting the profits of prominent US elites who possessed interests in northern Mexico.

Villa did have some capable Mexican representatives in the United States. One of his spokesmen, lawyer Federico González Garza, had been a close associate of Madero and had briefly served as governor of Mexico's Federal District. He wore dapper suits like Pani and Bonillas. He had overseen the inauguration of dredging works in the Gulf port of Frontera, but had not directly managed the project himself. He possessed no engineering or construction skills. The project was realized by workers under the direction of the North American Dredging Company and Mexican engineers Arturo Pani (Alberto's brother) and Manuel Urquidi (Francisco's brother).[72] Enrique Llorente was probably Villa's most influential representative and media operative in the United States. A lawyer, he was well educated and well spoken. But, like Francisco Urquidi, he spent most of his time propping of Villa's image and placating critics of policies proposed by Villa's camp that might damage US interests, instead of discussing reconstruction. This was a key difference.

Exploiting this difference was an important part of Constitutionalist diplomacy in the consequential years of 1915 and 1916. While the Wilson administration was debating which of the revolutionary factions would be best at reestablishing order and best for US interests, Wilson let it be known that the US policy of "watchful waiting" was over. If the vying powers could not come to terms, his government would ultimately back the force that it thought would best uphold liberal governance, respect religious freedom, carry out necessary agrarian reform, and expand education. Underlying this idyllic rhetoric, which the Wilson administration often contradicted in its Latin American interventions, was its drive to back the faction that would best benefit stability and US business. The Constitutionalists ultimately made the stronger case. Even though many Wilson officials cared little for Carranza's non-conciliatory nature and stubborn nationalism, he and the Constitutionalists were the dominant militant force by autumn 1915. They also had the clearest plans for national reconstruction and the most experts to carry them out.

In addition to military dominance, Carranza had a larger and more organized cadre of technocratic officials. They played on US proclivities for order, modernity, capitalism, and progressivism. It was people like these technocratic agents who influenced President Wilson's envoy DuVal West to conclude that the leading members of the Constitutionalists were of "a higher type" than their counterparts in Villa's camp. These same Constitutionalist agents swayed US muckrakers such as Lincoln Steffens, who had some influence on the Wilson administration. John Lind, a prominent Wilson envoy, had originally backed Villa but shifted his support to Carranza's forces because he believed that they were better educated and more similar to US progressives.[73] Steffens summed up this appeal: "I trust they [Carranza and his advisers] are like some of the best reformers we have ever known in this county [the United States]."[74]

Rolland, Bonillas, Pani, and Cabrera had studied US education systems, expanded communications in Mexico, managed railways, and written agrarian reform policies that were published in English and Spanish in the United States and Mexico. Villa, and Zapata for that matter, greatly appreciated education and possessed significant organizational capacities. However, while Villa's agents were desperately striving to salvage his image, Carranza sent fifty teachers to study US public school systems.[75] In October 1915, when the US secretary of state, Robert Lansing, had given word that the Wilson administration would back the Constitutionalists as the de facto government, he said that "the Carranza party is the only party in Mexico which possesses the essentials for recognition." It possesses military superiority, embraces liberal and progressive ideals, and better possesses "the material and moral capacity" to rule.[76] That the Constitutionalists possessed moral superiority was a highly questionable claim, though it was one that Carranza agents made regularly. But their superior technical expertise and territorial dominance were clear. Constitutionalist agents continued to reinforce this argument in subsequent years as they strove to consolidate their control of Mexico, preserve the peace between themselves and the US government, and maintain official recognition from the United States.

The difference between the Carranza and Villa camps was made clear by the Constitutionalist technocrats serving as agents in the United States. Their combination of progressive persuasions with technical expertise impressed important scholars, liberal middle-class professionals, and members of the US Progressive movement. Making up constituencies that largely supported and influenced the Wilson administration, this influence was not unsubstantial. Mexican engineers who served Carranza in diplomatic roles in the United States ultimately did a better job than Villa's representatives in selling themselves as the individuals most likely to reconstruct an orderly Mexico that would lend itself to increased development and a Mexico that would be more similar to the world that US progressives envisioned: clean, urban, fair, educated, liberal, and connected by modern infrastructure. In this way Mexican engineers during the revolutionary era not only shaped policies and the physical environment, but also the opinions of prominent citizens belonging to Mexico's northern neighbor, the most powerful nation in the Western Hemisphere, and the country outside of Mexico with the greatest capability to influence the outcome of the revolution.

CHAPTER 7

PUNITIVE ENGINEERING AND MILITARY MODERNIZATION

Reform, Revolution, and Reconstruction in Mexico and the United States, 1916–1924

JAYSON MAURICE PORTER

Anyhow you Motor Transport men, I am sure you will agree that all of us owe Francisco Villa a debt of gratitude; for he was the cause whereby the United States was somewhat prepared to meet the great problems confronting us on our entry into the World War and our subsequent participation in this struggle.

—**Colonel John A. Madden, "Thanks to Villa," 1921**

As hundreds of motor vehicles raced across the US–Mexican border into Chihuahua, Mexico, at the outset of the Punitive Expedition in March 1916, US officials predicted that the horseless age had finally arrived. Observers assumed US automobiles were destined to replace the cavalry, but they were incorrect. The introduction of the motorized military expedition did not immediately supplant the horse as a necessary element of military mobility, in part because of the lack of motor roads. After months of road building, US military officers realized that the combined application of horse and motor power made for a more sustainable cavalry than a new, purely mechanized service. They used words like "lessons," "experiments," and the "scientific method" to describe successes and failures, which would continue to inspire national defense projects in the United States and Mexico. This punitive engineering, or military-related infrastructure, was key to maintaining national stability in the wake of the Mexican Revolution.

More than the first large-scale motorized campaign in military history, the Punitive Expedition was the first campaign to require a motorized road. This chapter tracks the technological and organizational lessons that US and Mexican leaders learned from expedition and revolution.[1]

Military-related infrastructure has a storied history in how political institutions consolidate and stabilize power and responsibility, or state building. Political bodies, such as kingdoms, empires, and metropoles, have historically relied on military roads, forts, and arms to define, expand, and maintain political momentum. Because of the military's proximity to science and engineering, states and empires called on the military to create infrastructures like bridges, irrigation systems, and naval ships. Long before US president Eisenhower coined the term "military-industrial complex" after World War II, military-related infrastructure helped sustain state-industry relations. Even after the rise of civil engineering in the late nineteenth century, military engineers held a privileged position as state builders.

This was especially true along the US–Mexican border where both nations defined and remade themselves, and did so with the military. From wars against Indigenous peoples to wars between nations and political revolution, the borderlands served as a site where state building took place through punitive engineering. Professionalizing civil engineers began to eclipse military engineers in twentieth-century urban development, but real and imagined threats to national security inspired military action in the borderlands. But even along the border, this state building looked different in each country. State building in both nations had transnational impulses for order and progress that ignited during the Mexican Revolution, but while Mexican punitive engineering sought domestic power, US leaders geared their punitive engineering toward both domestic and foreign power.

The Punitive Expedition, which lasted from March 1916 to February 1917, influenced Mexican state formation and US empire building for decades to come. Seemingly overnight this military episode helped reveal the social, political, and transnational implications of military-industrial relations when those relations were increasingly central to state building. What started as a hunt for Mexican general Francisco "Pancho" Villa after his assault on Columbus, New Mexico, on March 9, 1916, became a major engineering challenge. The US military had built roads abroad earlier, but because the expedition was experimental, close to home, and highly expensive, those roads affected the new military and civil road construction. Chasing Villa in motor vehicles required building new roads in new ways, which helped create a broad sense of urgency for both semi-motorized militaries and domestic military roads for US and Mexican national defense. For the Mexican armed forces, the militarization of the border in 1916 first helped Villa's enemies secure political and technological victories in the

Mexican Revolution and then influenced how they maintained political power militarily.

This widely felt anxiety to catch Villa inspired new engineering standards. Journalists, politicians, and public figures observed the military's engineering challenges, failures, and accomplishments.[2] The experimental exercise they witnessed tested roads, horses, and motor vehicles as it did soldiers, state formation, and policies. Federal, military, and regional authorities perceived the expedition as a national engineering challenge, which encouraged them to collaborate in novel ways. Less than three months into the Punitive Expedition, the US Congress passed the Federal Aid Road Act (under the National Defense Act), which offered the first federal distribution system of funds for states on the grounds of commerce and national defense. With the help of politicians and businessmen, soldiers built emergency cantonments, or army bases, and expeditionary roads that became the mobilization model the entire US military used to prepare recruits for the Western Front in World War I. Despite these linkages, historians seldom credit the Mexican Revolution as a proximate cause of the 1916 National Defense Act and its centerpiece Federal Aid Road Act.

Histories of the US and Mexican militaries seldom acknowledge the influence of the Mexican Revolution as a spur to technological advance in general. Historians of the United States tend to credit World War I for the expansion and technological dominance of the US military, whereas historians of Mexico often criticize postrevolutionary military officials as predatory obstacles to state building and development. This essay refutes neither point but argues that the expedition and revolution were engineering challenges with technological and state-building solutions in both countries. This is especially true in Mexico, where military engineering projects were essential to state building in the wake of the revolution. Even before the National Road Commission in 1925, roads were already among the most foundational reconstruction efforts.[3] These military roads and other forms of punitive engineering helped the Álvaro Obregón administration defeat its last major armed struggle, the De la Huerta Rebellion of 1923–1924. Interpreting both the Punitive Expedition and the Mexican Revolution as "contact zones" recasts military engineering projects as state-building solutions imbued with transnational cooperation and national defense goals.[4] US and Mexican engineers forged new transnational, state-building, and technological solutions to military, civil, and social engineering challenges.

US and Mexican engineers promoted new methods of industry, sanitation, agriculture, and civil engineering. In Mexico engineers connected industries such as construction, armaments, and agriculture to numerous departments of the postrevolutionary government, namely the Departamento de Salud Pública (Department of Public Health, DSP), the Secretaria de Comunicaciones y Obras Públicas (Secretary of Communications and Public Works, SCOP), and the Secretaria de Agricultura y Fomento (Secretary of Agriculture and Development, SAF). In both countries modernizing engineers helped establish early and smaller-scale military-industrial complexes. Much of this development relied on transnational cooperation. While transnational narratives of this period tend to underscore hostility and distrust in US-Mexican relations, below is an analysis of cooperation between the militaries.[5] By tracking the two militaries' development of roads, cantonments, and other conduits of defense technologies during revolution and expedition, this chapter argues that US and Mexican military engineers influenced state building by producing, using, and sharing technological knowledge.[6] The term cantonment, unusual in Mexican history, was the category used by military officials to describe what were effectively warehouse bases. Unlike those in imperial India, these cantonments varied in size and components but generally had barracks for soldiers, storage for equipment and arms, and hippodromes for horses and recreation. Both Pershing's 1916 Expedition and the revolution itself accelerated and altered technology transfer, in turn impacting interwar North American state building and technological advancement for years to come.

After punitive engineering, cooperative technology transfer and military construction helped Mexican revolutionaries establish control during the reconstruction period. More than purchasing arms, auto parts, and road-building machinery, revolutionaries also spearheaded engineering projects related to irrigation canals, road construction, and agribusiness with US support. Herein lies another argument: technology transfer as cooperation demonstrates informal recognition of Mexican revolutionaries as legitimate state authorities years before the Bucareli Accords reestablished formal US recognition of maintained commercial and institutional relations with Mexican military officials. As with the Rockefeller Foundation's joint Yellow Fever Campaign with the Mexican military (1921–1923), for example, military engineering campaigns helped reestablish US-Mexican relations before Bucareli. The exchange of ideas and technologies demonstrates how the circulation of military-related technologies buttressed the technical foundations of the postrevolutionary state before the formal modernization of the Mexican military in 1925.[7]

FROM MILITARY REFORM TO ARMED REVOLUTION, 1865–1916

The Punitive Expedition may have accelerated the rate of US–Mexican military technology transfer, but it did not begin that exchange. Like the US Republicans after the Civil War, Mexican liberals after 1867 also wanted a stronger, more organized military and set in motion a series of centralizing military reforms that aimed to improve academia by adding scientific discipline and engineering education, but continuing political chaos and economic troubles hamstrung such initiatives.[8] In the late nineteenth century the expanding US and Mexican armies did more than keep political conservatives at bay; as state-building tools, they also helped expand territorial control in the frontier. Military reform in both nations was increasingly tethered to violent acts of state building, namely wresting land from Indigenous peoples for settlers instead. Punitive expeditions against Indigenous peoples aimed to destroy with one hand and build with the other. Mapping equipment and munitions served the same goal: dispossession. Toward this end military engineers were central to this social and spatial reorganization of rural land from Texas to Tlaxcala, because manifest destiny and *mestizaje* were both military campaigns against people and place. Before the Punitive Expedition of 1916, US officials used the term "punitive expedition" to justify their attacks as forms of vigilante justice and revenge. Mexican officials did not readily use the term, but they had similar practices and even occasionally participated in US punitive expeditions, such as those against the Apaches in the 1880s. Despite this shared history of settler colonialism, the Punitive Expedition became a proper noun at the expense of the Mexican people.

During the Porfirian era, US and Mexican militaries also interacted at technical schools. While some Mexican officers spoke choppy English, and most US officers butchered their Spanish, everyone spoke military. The Colegio Militar was an early site of military communication.[9] It was also a unique place to learn engineering. Before the rise of civil engineering professionals in the early twentieth century, only mining and military schools offered instruction in engineering. With a curriculum varying little from curricula at other leading military academies, the Colegio Militar received students from Germany, Japan, and the United States.[10] Mexican students, especially engineers, traveled abroad too. For instance, engineers Rodolfo Casillas and Victor Hernández Covarrubias went to Fort Riley to study explosives and artillery.[11] In addition to taking classes, both men shared ideas and sold Mexican-made arms to technical schools as they traveled.[12] They likely sold General Mondragón's rifle, which was the world's first semiautomatic rifle. These "military universities" were places where officials experimented with new cavalry, light infantry, and military engineering techniques. Engineers also tested techniques in the field as the US and Mexican militaries increasingly practiced these techniques against Native Americans in the 1870s and 1880s. Designed to punish enemy forces that officials pathologized as dangerous or harmful to order and progress, experimentation on these expeditions reinforced statecraft, empire, and progressive liberalism.[13] Historians of the Porfirian army describe experiments in artillery and counterinsurgent tactics, and similarly, during and after the revolution, new armed forces in Mexico would develop, test, and share new military technologies and techniques.[14]

The Mexican Revolution transformed the military into the nation's principal state builder. Like the regime it served, the brittle Porfirian army broke apart in less than one hundred days. But the its officers cohered around Victoriano Huerta, as Madero learned to his regret. It took a bloody, yearlong war to defeat the Huertistas, but many Porfirian military engineers outlasted the regime to become essential organizers in the armies of the Zapatistas, Villistas, and especially Carrancistas. These military engineers provide a unit of analysis to measure the structure and political culture of each faction. In short, more engineers resulted in more provisions, more organization, and more victories. Villa famously ignored the advice of engineer Felipe Ángeles. By contrast, Carranza's engineers helped his army secure oil, steel, and timber industries, which in turn financed and supplied the Carrancista army, helped defeat Zapatistas, and pushed Villistas to the edge. From this position—so far from the central power, so close to the United States—Villa decided to cross into New Mexico, attack Columbus, and murder eighteen US citizens.[15]

PUNITIVE EXPERIMENTATION, 1916–1917

The Columbus attack marked a turning point in U.S-Mexican military relations. First, it required the expansion of the US armed forces. The US Army numbered fewer than 107,000 troops, of whom 25,000 lived abroad. By contrast, as the secretary of war in 1916, Álvaro Obregón commanded more than 200,000 enlisted men. However, chasing Villa helped the United States to expand its army. The peak of the Punitive Expedition stationed 1 of every 11 US soldiers in Mexico. Expedition leaders John J. Pershing and Frederick Funston came to Mexico with experience in the Philippines and Cuba. Engineers like Captain Ernest Graves also crossed the Pacific to Chihuahua, where the US–Mexican border presented him with new problems that required novel solutions. For the US Army and National Guard, the Punitive Expedition was a series of mass mobilizations with an additional challenge of supply chain management, which involved hundreds of trucks to support cavalry expeditions and several warehouse bases to store materials. By July 1916 over 100,000 guardsmen "protected" the border, with another 40,000 in state mobilization camps.[16]

Even though the Punitive Expedition failed to capture Villa, it did catch the attention of US politicians and the public. The expedition went beyond military experimentation—it was an object lesson that motivated US state building. The expedition was a highly visible national emergency, a daily headline in many newspapers, from which observers read about US officials' struggle organizing supplies and then mobilizing forces to the border. Various companies and businesses jumped at the opportunity to help and sent equipment for the effort. Chicago business owners, for example, raised $25,000 to purchase twenty machine guns to accompany the Illinois Guardsmen; Goodyear Akron gifted the Ohio Guard an observation balloon;[17] Ford's and Packard's employees drove hundreds of vehicles to the border.[18]

Engineering challenges began immediately. For nearly a week the army waited for the authority to cross the border. From the expedition's nerve center in Fort Sam Houston, San Antonio, organizers sent horses, motor vehicles, and men to temporary bases in Texas and New Mexico. Military officers understood that the intervention would require wartime camps and reliable lines of supply and communication. The problem of supply management in particular was a looming concern that drew competing solutions. Funston saw cavalry as the military's best bet, but Pershing foresaw a foraging problem in the barren Chihuahua landscape and recommended motorized operation.[19] They compromised on a semi-motorized operation, and by the expedition's end, the US military had built 157 miles of road, repaired another 224 miles, put up 677 miles of telegraph lines, and used over eight thousand horses and mules in Chihuahua.[20] President Wilson ultimately discouraged Generals Pershing and Funston from occupying any Mexican towns, to which Funston happily responded, "Anyway, it will get us an army."[21]

At 3:17 p.m. on Thursday, March 16, the US military crossed the border into Chihuahua with an initial force of 4,800 men.[22] The forces traveled an average of 18.8 miles a day until they reached Namiquipa in late March and built their forward operating base 243 miles south of Columbus.[23] Between Columbus and Namiquipa, Quartermaster Colonel John Madden established Colonia Dublan to store supplies for an increasing number of soldiers. In less than three weeks, the expeditionary forces expanded by nearly 50 percent to 6,675 troops.[24] Their progress became slower thanks to the fact that the Mexican revolutionary leaders did not grant US authorities access to Mexican railroads. Instead Obregón gave Pershing and Hugh I. Scott "Obregón passes" to officers, which functioned like visas. Without the authority to occupy towns or use railroads, the US expedition for Pancho Villa quickly devolved into an experiment in supply chain management with road building as the highest priority.[25]

At first the narrow dirt roads stood up to the wear of the cavalry and motor vehicles.[26] However, with seventeen automotive companies transporting over 78 tons of cargo daily, truck convoys soon dredged impassable ruts into the Chihuahua countryside.[27] With only 65 automobiles and 105 trucks in the US Army before the expedition, officials were simply unfamiliar with putting so much wear on roads.[28] New ones were vital. Aware of the transport supply problems and fearing the rainy season, engineers stationed in Columbus began to request road-building equipment in late March and started pioneering practicable routes in April.[29] Officials were initially concerned with too much water. However, without spring showers, they would lack the water needed for construction. With no water, dust ruined machinery, perishables, and roads. Twelve truck-mounted machine shops and several water-tank trucks could not carry enough.[30] Such dry weather problems in Chihuahua were quite different from the rain problems military road builders had faced in the Caribbean and the Philippines.[31] In northern Mexico dust was the "consistency of flour," and it covered and contaminated everything, especially water for drinking and engineering. In this arid environment, before road building started and vehicles chased Villa, soldiers searched desperately for potential wells while engineers reconnoitered for supply routes.[32]

Plank roads would have offered an alternative had there been more available timber. But consequently, the first "decided departure from the usual road-making principles" was the sunken or shaved road for the dry season. The region's alkali dust particles could not be packed, molded, or fixed into a graded road, so engineers—or more accurately Mexican laborers and black soldiers—resorted to "shaving" the road surface down to the bedrock.[33] Because sunken roads needed constant shaving, they demanded more labor and new scales of maintenance. Builders relied on five classes of machines: tractors, graders, rollers, scrapers, and drags.[34] Soldiers experimented with most machines in this landscape: some techniques became standard; others fell out of use. Notably, officials gave soldiers Colt 45 pistols to test in Mexico, but the test backfired, and soldiers who tried to use the weapons from horseback were injured when the weapons recoiled, leaving them with bruised and bloody hands. As road-building machinery tractors became as necessary as trucks. Only graders longer than twelve feet worked; soldiers discarded most drags due to inutility.[35] As for automobiles, new adjustments, such as khaki paint (to match the dust) and higher chassis, became new standards.[36] Observing the military use of motor vehicles also taught automotive companies much about the effects of dust on governors, wheels, and transmissions, and like road builders in Chihuahua, they frantically sought the best engineering practices.[37]

Punitive engineering in this environment demanded more than road building. Roads did not directly encounter, much less counter, insurgents; instead they supplied soldiers with the weapons and rations to chase Villa off-road. Officials marshaled those supplies in bases such as Colonia Dublan. The military base in Columbus, New Mexico, in particular went through a "crash program in mechanization" after Villa's assault.[38] Colonia Dublan had storage space, offices, and a hospital on nearly two acres at the

Fig. 7.1. "U.S. Army trucks on way to Mex[ico]." *Source*: James A. O'Connor, "Road Work in Mexico with the Punitive Expedition," *Professional Memoir* (*Military Engineer*), December 1917, 328.

Fig. 7.2. "The compacting process was a slow and laborious one." *Source*: James A. O'Connor, "Road Work in Mexico with the Punitive Expedition," *Professional Memoir* (*Military Engineer*), December 1917, 337.

expedition's peak.[39] Across Texas the National Guard built other border-land cantonments.[40] Constructing camps and their connecting roads was the first and only objective for many militiamen. Within months a string of camps began to reshape the borderlands economy. From San Antonio, the center of operations, to Arizona, military officials increasingly hired citizens for these engineering developments, and in turn, base construc-

The dirt roads were good for the first few days.

Fig. 7.3. US Army trucks in the Chihuahua desert: "The dirt roads were good for the first few days." *Source*: James A. O'Connor, "Road Work in Mexico with the Punitive Expedition," *Professional Memoir* (*Military Engineer*), December 1917, 327.

The sunken or "shaved" road.

Fig. 7.4. "The sunken or 'shaved' road" for the US Punitive Expedition. *Source*: James A. O'Connor, "Road Work in Mexico with the Punitive Expedition," *Professional Memoir* (*Military Engineer*), December 1917, 339.

tion gained a reputation for providing job security and ensuring national defense. After seeing the economic boom in Columbus and San Antonio, citizens in other cities, such as El Paso and Deming, requested permanent military establishments for economic reasons.[41] The Punitive Expedition, however, absorbed most of the military's attention and resources.

Those needs only increased with the arrival of the rainy season. Dry season construction was labor intensive, but US Army leaders anticipated even more labor shortages once too much water became the problem. Their solution was Mexican labor. With new authority to hire Mexican laborers, the army began preparing for the wet season and maintaining sunken roads for the dry season at the same time. Access to Mexican laborers also gave Pershing the power to create and train a local militia to assist in the hunt for Villa. Much to the surprise of Namiquipa residents, Pershing's troops exchanged gold coins for hired labor, cattle, and forage.[42] Given the high cost of construction, the practice of localized road building gained traction during the wet season. The idea behind localized construction was simple: roads are best when they are built by small detachments imbued with "proprietary interest in how well their roads were to be built."[43] This practice of "localization" was a major reason why army engineer Ernest Graves believed that "the punitive project embodied perfect military road-building." Graves argued that regardless of local ecology and topography, local engineering was "more a matter of organization than of special technical knowledge."[44]

Not everyone agreed. Many engineers appreciated these lessons in hindsight, but in situ expedition engineers argued incessantly over road-building practices during the expedition. Most notably they disagreed over how to use rainwater. Pershing ultimately removed the chief engineer for his unwillingness to try to construct graded roads with rainwater. This drastic action forced the engineering corps to reorganize into the Second Regiment of Engineers, which was the first time the US Army created a regiment outside of US territory.[45] Graves would lead the unit from western Mexico to the European Western Front, where they built roads in France in 1918 and 1919. The Army Quartermaster Corps and the Society of Automotive Engineers also co-engineered and then sent thousands of Liberty trucks (class-B standardized vehicles tested in Mexico) to Europe, which further embodied this technology transfer.[46] With punitive engineering, "Atlantic Crossings" went from Mexico to the United States and then Europe, not the other way around.[47]

Another civilian by-product of the Punitive Expedition was the Good Roads movement.[48] As the expedition accelerated, millions of US citizens came to associate road building with national security. Articles with titles like "Good Roads Vital Factor in National Defense Campaign" featured in major newspapers. Difficulties experienced near the border resonated with citizens around the country who also encountered bad roads. On July 11, 1916, Congress passed the Federal Aid Road Act in conjunction with the National Defense Act, and overnight the US Bureau of Public Roads became the national highway department around which individual state highway departments would revolve.[49] With the National Defense Act and its Federal Aid Road Act in motion, the emerging road-building frenzy drew industry and the military together by assisting states to fund and build military-grade motor roads. In support of Good Roads, Theodore Roosevelt and other military politicians even held a diplomatic conference on the Mexican border where they proclaimed, "Highways were a military asset, almost as vital a factor as a large standing army."[50] Good roads connected US businessmen, politicians, and consumers in new ways.[51] This was especially true in the automotive industry, where the adoption of military-grade standards, such as new wheel weight, chassis height, and axle width, had commercial ramifications. Good roads were true in other cases as well. Hewitt Callender, a former truck master of the Pershing expedition, went to speak at the Los Angeles Transportation Association; the Dixie Highway Association changed their good-roads promotion model from tourism to national defense; towns continued making bids for constructing bases and their concomitant roads.[52]

Military-civilian road construction proved to be the expedition's farthest-reaching legacy. When General Pershing returned to Washington in early 1917, he commissioned a recent West Point graduate to design a national system of military highways.[53] The lieutenant's name was Dwight D. Eisenhower, and he produced a map of 75,000 miles of strategically placed roads that became known as the "Pershing Map." Road builders continued to refer to this map until the 1950s.[54] Pershing did not live to see his influence on the US Interstate Highway System, but he did know that the experiment was a driving force behind the widespread construction of military bases in the US Preparedness movement.[55] Militarily speaking, the expedition was insipid; technologically, however, it was inspirational.

When the United States entered the Great War two months after the Punitive Expedition, they expanded cantonment construction from along the border to across the nation. As they had in the Punitive Expedition, these warehouse bases served as cooperative, economic, and contact zones in which good roads were built in the United States and eventually Mexico. According to *Scientific American*, cantonments were critical to the transition from the Punitive Expedition to the Western Front because "the housing of the new army . . . [was] the greatest problem ever tackled by any government."[56] This hyperbole notwithstanding, from June to September 1917 the US Government reallocated millions of dollars to build sixteen cantonments like those from the borderlands.[57] Each cantonment was two miles long, equipped with "water supply, sanitation, heating, lighting, and power." Moreover, they doled out salaries to hundreds of workers and soldiers.[58] Each cantonment contained dozens of buildings, 400 miles of electric wiring, up to 25 miles of road, and dozens of horses for each automobile. Most cantonments also manufactured products: some made explosives and small arms; one plant flirted with making chlorine gas.[59] Cantonments absorbed tons of products too. For lumber alone, cantonment builders purchased wood from 190 different lumber mills.[60]

Successful cantonment construction did not guarantee good road construction. Despite the initial boom, federal aid did not save US road builders from major construction lulls from 1918 to 1920. The 1916 Act established a distribution system for federal aid to state roads but grossly underestimated the construction costs and contingencies. As early as the following year, observers noted that the United States seemed to contain "48 state highway departments with 48 standards of road building." Moreover, less than 1 percent of the mileage was suitable for "continuous use."[61] Military-civilian contracts in 1918 revitalized the US Bureau of Public Roads and the Committee on Emergency Construction once road construction machinery returned to the United States after the war. However, redistributed machinery could help only for so long.[62] In 1920 US road builders failed to construct 75 percent of the roads they planned to build that year. With lessons from 1916 Congressional passage of the Federal Aid Road Act of 1921 ushered in the golden age of US road building.

MILITARY REVOLUTION TO MILITARY REFORM, 1917–1924

In the wake of the revolution, military modernization was also essential to Mexico's reconstruction. While no historical records describe this process in full detail, the Fideicomiso y Archivos de Plutarco Elías Calles y Fernando Torreblanca houses important documents on the topic, of which the US military attachés' reports are especially illuminating.[63] With expressed interests in politics, industry, and the residual social unrest, military attachés made interesting but incomplete observations, which they sent to commanding officers, stateside authorities, and the US Embassy in Mexico City. With the completion of the Panama Canal; military occupations in Haiti, Nicaragua, Cuba, and the Dominican Republic; and the rise of dollar diplomacy in the Caribbean basin, the number of US-directed engineering projects in Latin America dramatically increased during the interwar period. Road projects were commonplace, and as was the case with railroads, Mexico's market for roads was of unique importance to US business.[64] In the wake of the Mexican Revolution, US military officials, construction companies, and business owners visited Mexico with commercial dreams that demanded motor roads.[65] It is significant that from 1920 to 1923 road building and other engineering projects connected US-Mexican military relations even as the US government formally refused to acknowledge the legitimacy of the Obregón administration. Starting with Henry Ford and Herbert Hoover in 1918, each year only brought more investors in efforts spearheaded by Mexican military leaders. Attachés such as George M. Russell, R. M. Campbell, and Edward Stone were keen to uncover and disclose the inner workings of Mexico they envisioned as "America's Market."

In the reconstruction phase of the revolution, President Álvaro Obregón's administration strengthened the state through new military connections to industry and civilian sectors of society. Military-related projects in these contact zones helped establish political stability in the wake of the revolution. Military control of wireless radio helped Obregón and his supporters facilitate the revolt and maintain power in Mexico.[66] Similarly, road building and military establishments helped the Obregón administration centralize federal authority, defeat the De la Huerta Rebellion, and advance economic development before 1924.[67]

Obregón's prerevolutionary experience as a municipal president, chickpea farmer, and amateur inventor gave him civil, social, and creative insights that supported his military prowess. He navigated politics and business-like battles—victoriously. Even after adopting his new civilian role as president of the Republic, military concerns informed his governance. Obregón extended military officials material gain for political subordination. However, without the official US recognition and loans extended to Carranza, Obregón's government initially struggled to pay its military, which absorbed 61 percent of the national budget in 1920. To curb the threat of an unruly army, which could test Obregón's authority, the president turned to military professionalization and economic development. In both processes of punitive engineering, military engineers were key state builders during the administration's three years without official US recognition and economic support.

Generals-turned-politicos had good reason to love engineers. From the administration's beginning, construction projects created jobs for educated ex-soldiers. It was helpful that national leaders cooperated across government branches. By early 1921 future presidents Plutarco Elías Calles and Pascual Ortiz Rubio—the latter an engineer himself—were named ministers to the Secretaría de Guerra y Marina (SGM) and to SCOP, respectively. After Obregón removed Jacinto B. Treviño from duties overseeing national demobilization efforts, the president "created nineteen special work battalions for roadbuilding, irrigation development, and railroad and telegraph repair."[68] Congress helped accelerate demobilization by passing the Law of the First Reserve, which incorporated 452 generals, 2,290 colonels, and 8,318 junior officers into a reserve and provided the discharge of all their enlisted men.[69] With a stronger reserve, a cooperative cohort of ministers, and record oil revenues, the administration coupled demobilization efforts and engineering projects to reduce the military budget by 39 percent that year.[70] More than alternative employment for officers and their discharged soldiers, the reserve allowed its members to retain the prestige of being men in uniform. Military-built roads, camps, and colonies were part of the material exchange through which Obregón demobilized soldiers.[71]

Military state builders of the time had ambitious plans. They would not build cantonments for a few years, but as early as 1920 officials authorized soldiers to construct military establishments, such as colonies, encampments, and barracks. Leaders also hoped for moral uplift, which required improving the military academy and placing barracks outside the city. After reopening a Colegio Militar that many hoped would become "the West Point of Mexico as it was in the days of Porfirio Díaz," officials turned their attention to army barracks.[72] To reduce soldiers' alcohol and marijuana use and stifle military crimes, generals socially engineered new barracks at a distance from the bad influence and temptations of Mexico City. SGM and SCOP officials decided to sell all the quarters in the city for seventeen million pesos to help fund twenty-seven new barracks in Mexico City's outskirts in 1922.[73]

Military-related development extended well beyond the Valley of Mexico, and agricultural colonies were popular outside the capital. In Tamaulipas and Nuevo León the government organized six small agricultural colonies averaging 7,000 acres for a total of 427 families, who received 80,000 pesos of federal aid for land, tools, seed, and other needs.[74] The First Army Reserve helped relocate entire battalions of discharged men to agricultural colonies in Chiapas, Chihuahua, San Luis Potosí, Puebla, and Michoacán.[75] In the First Army Reserve in Michoacán, Obregón gave general and engineer Enrique Estrada 3 million pesos to establish military-agricultural colonies. In Coahuila General Joaquín Amaro collected funds to organize "a great cavalry camp." Amaro's cavalry camp was so "modern" that US attaché R. M. Campbell noted, "Such a large, complete and orderly encampment has not been seen by [Russell] for many years."[76] The archives are silent on many details about the earliest reconstruction agendas, but it is clear that punitive engineering projects stimulated the economy by employing soldiers.[77]

While the War Department focused on new barracks, colonies, and camps, other ministries cooperated across bureaucratic lines to initiate the First National Congress on Roads in September 1921. University leaders and state bureaucrats planned the congress to coincide with the national centennial celebrations in September. The decline in railroad revenues since 1918, combined with the natural shortage of coal reserves, added to road builders' sense of urgency.[78] Given the symbolic and economic stakes, the coordination necessary to find a venue, invite the best participants, and make proposals to debate required military, civil, and social engineers. Many civil engineers organized this congress, but as this chapter argues, working with military officials was a hallmark of the period.[79]

Engineer and Subsecretary of Public Works Faustino Roel set the commission's agenda. After he secured the venue—the auditorium and assembly hall of the National School of Engineers—Roel sent invitations to the governors of each state, the municipal presidents of every capital, and the directors and presidents of universities.[80] The engineer also notified leaders in the Society of Engineers, various chambers of commerce, sports clubs, and interest groups he expected to help finance roads.[81] The goal was to bring business leaders, politicians, and academics together. With a venue and participants lined up, who would build the initial roads? The Mexican Congress needed roads to compare and contrast, and the military provided much of the labor and command to build them. Four months before its first meeting in September, the Congress appropriated 30,000 pesos for a borderlands route from Ambos Nogales to Tubutama, Sonora, and long-standing projects from Mexico City to Acapulco and Juárez.[82] Within months, US attachés noted the deployment of military engineering commissions to map potential routes across the country and soldiers and peasants eventually doing the roadwork in both northwestern and southeastern states.[83]

Road construction could engender resistance in many forms because roads often ran through Indigenous lands and absorbed a disproportionate amount of government resources. In a Yaqui community between Nogales and Tubutama, Sonora, locals discovered the bodies of a civil engineer and five road workers.[84] Officials blamed the Yaqui but made no arrests. Roadwork continued, albeit slowly. In Mexico City protesters manifested as another hurdle for road construction. Marching against the "Army Avenue" running from the Colegio Militar through Mexico City, protesters asked why the government should build that single road at the cost of ten schools.[85] As Matthew Vitz's chapter in this volume shows for Mexico City, communities occasionally resisted engineering projects with local solutions.

In other communities, locals supported road construction, but they hoped to influence where the roads went. For example, in Guerrero residents of Taxco wrote to President Obregón with a road proposal regarding the nearby Mexico City–Acapulco highway in progress. The locals insisted that a road deviating to the "wonderful caves of Cachuamilpa" could help the region's paralyzed mining, agricultural, and commercial sectors. Referring to an 1882 engineering report, they argued that even though a route to Cachuamilpa was seventeen kilometers longer, it could cover costs by taking advantage of an old bridge over the Huajintlán River. In addition the residents of Taxco and surrounding towns offered their labor.[86] Forced by economic constraints, the residents and neighbors of Taxco responded with a sophisticated countercyclical strategy. Through intraregional and region-state cooperation, these residents hoped to spend their way out of economic hardship, though the sources do not mention if they succeeded.

Local agendas mattered, but military concerns still dominated fiscal agendas. The 1922 budget promised a 33 percent reduction in military expenses with further consolidation efforts. Austerity required teamwork.[87] When the government eliminated the Aviation Branch of SCOP, its personnel transferred to the Aviation Branch of the War Department.[88] Austerity could also be unilateral and unforgiving; the army reorganized the Cavalry Branch without any lieutenant colonels and many of its captains to save 540,000 pesos a month.[89] Occasionally financial constraints encouraged institutions to make new branches, such as a corps to handle army supplies and transportation, a savings bank for soldiers, and the Inspection Commission of the Army.[90] In another example, the Artillery Branch absorbed several arms factories in Mexico City and purchased a cylindrical barreling patent to make better rifles.[91] The Artillery Branch saved the military 3.5 million pesos by manufacturing and refurbishing its own arms. But not even this could make ends meet. These changes reduced the military budget by 39 percent in 1922, but that was not enough to avoid the financial discontinuation of many road construction projects.[92]

Like US road builders before the 1921 Federal Aid Road Act, Mexican engineers had navigated through a national construction lull from 1922 to 1923 with local solutions and materials. Interest clubs and chambers of commerce were especially eager to assist with and even initiate roadwork.[93] For instance, in early 1922 the Automobile Club and the Chamber of Commerce of Coahuila's Laguna region inaugurated an earthbound macadamized motor road comprised of local limestone. In Monterrey and several cities in Veracruz, members of various clubs used oil export duties and repurposed local cement to fund public works projects. It was also common for mining and oil companies to build roads to transport or use mined materials.[94] Given the underutilization of intermediate goods industries during the Porfirian era, the integration of local materials into public works helped feed the reconstruction economy. Cement, steel, and dynamite factories that lost revenue due to Porfirian under-capacity production now witnessed more profitable years in Mexico's reconstruction because of the domestic market in public works.[95] Even though such local efforts could not support national ambitions, they were central to regional development in the borderlands.

Along the US–Mexican border, local-level cooperation often assumed transnational forms. The border economy offered more than contraband in the Prohibition Era; it also was a growing center for purchasing military-related products. Horses, auto parts, guns, and road-building machinery all attracted Mexican generals to the region. In fact, even in this motor age the horse trade made for bustling business. This was especially true because Mexico's military needed to replenish a horse population devastated by the revolution.[96] Luckily for horse buyers and sellers, San Antonio in 1921 was still the equestrian paradise it had been during the Punitive Expedition. When Joaquín Amaro authorized the commission to travel to Alamo City, they purchased one thousand horses.[97] San Antonio's International Engineering Company made contracts with other Mexican military commissions for more than ten thousand horses at a time. And when the company could not acquire the horses from breeders around San Antonio, they sent their engineers to El Paso.[98] One former US military captain, who worked privately as a horse merchant based in Texas, traveled as far as Wyoming and Montana to find horses for Mexican forces.[99]

Years before official US recognition of the Obregón administration, US politicians and businessmen in the Southwest extended Mexico's commercial recognition. Regional icons, such as William P. Hobby, ex-governor of Texas and personal friend of Obregón, were leading advocates for regional recognition, or de facto recognition, between bordering US and Mexican states.[100] By August 1921 thirty-three cities in Texas alone maintained Mexican Honorary Commissions to facilitate transborder trade.[101] Mexican generals sent their consuls to El Paso, San Antonio, and Nogales to buy automobiles and auto parts.[102] On occasion *militares* also purchased guns and road-building equipment.[103] Sometimes officers sought help from US officials in border towns and cities. Notably, in January 1922 the Mexican consul in Nogales helped General Angel Flores acquire a steam shovel for a 750-mile highway his forces were building from Nogales to Nayarit.[104] In this tomato-growing region between the cities, exporters and importers valued smooth motor roads to transport the fragile fruit safely. Once the road reached Culiacán, Flores managed to garner the municipal and federal support necessary to contract the US road builder Bill Goode to help complete the job.[105]

As a result of rapid agricultural and urban development in the 1920s arid US Southwest, US engineers became increasingly interested in northern Mexican waterways like the Colorado and Yaqui Rivers. When securing water sources to irrigate agricultural lands, US officials often sought help from their Mexican counterparts. The Colorado River was an early site of contention. In contrast to Flores's advantageous arrangement with officials in Douglas, Nogales, and Tucson, Arizona, the new governor of the northern district of Northern Territory of Baja California, José Inocente Lugo, had a testier relationship with officials from California and Arizona, chiefly the result of conflicts regarding water issues. Lugo sought socioeconomic reform through new roads and irrigation projects, but the governor was equally wary of US intentions and reluctantly agreed to represent Mexico in a commission on the Colorado River.[106] Since overuse in the US southwest adversely affected the water supply of northern Mexico, Mexican officials tried to avoid partnerships that benefited the United States at their expense.

When the Yuma Chamber of Commerce invited Lugo to visit and discuss irrigation, the governor was unsatisfied.[107] In San Diego next, Lugo met with Colonel Ed Fletcher and Mr. Darlington, the president of the Commission of Asphalt Roads in the State of California. The Americans courted Lugo with a road-building proposal the governor found unfavorable to Mexican interests.[108] In both cases tensions were highest in matters concerning resource allocation and rights directly on the border. Transnational cooperation was seldom smooth, and even though at its best in the borderlands, US and Mexican officials tended to cooperate better on either side of the boundary than on the border itself.

It was among military officials that transnational cooperation was generally at its most amicable. Rodolfo Casillas and Victor H. Covarrubias's four-month tour of US military academies, cantonments, and industries offers a good example. Before becoming engineering professors at Colegio Militar, both career military officials studied and worked in US military facilities during the Porfiriato. Casillas spent three years at Fort Riley, Kansas, where the 1922 tour began. The tour continued to Washington, DC, Annapolis, West Point, and New York City. Accompanied by US military officials, the consulting engineer of National Railways of Mexico, and a government architect, they visited cavalry schools, bases, and arsenals. Base knowledge transfer aside, the tour did not directly promote the military construction projects or economic development in Mexico as officials had anticipated.[109] But openly accepting US military support at the national level proved difficult. Obregón experienced this personally in 1923 when the national newspaper, *Excélsior*, ridiculed him for his friendship with and visits from a US colonel, McNab.[110] Transforming the Mexican military into a state-building force demanded more than collaboration between military officials and de facto recognition from business owners.

If neither regional nor military transnational cooperation could revive military construction efforts by 1923, did formal US recognition established under the Bucareli Accords of that same year rescue Mexico from its military? It did not: neither US recognition nor the loan agreement guaranteed Mexico the economic stability that both countries desired. The recognition boom of the Bucareli Accords was short-lived, but the diplomatic agreement streamlined a few important state-building projects.[111] However, formal recognition did reignite the development of pre-Bucareli projects.[112] In this sense military reconstruction and foreign finance were mutually reinforcing, and therefore neither technological determinism nor economic dependency was fully at play in postrevolutionary Mexico. Of these pre-Bucareli projects, the most important were the twenty-three *sueltos*, or detachments, built to support a system of permanent and temporary stations to deploy the sueltos quickly and strategically. Soon enough the sueltos that attachés described as "flying columns," would play an important role in harnessing government control when the last nationwide postrevolutionary rebellion ignited in December 1923.

As the pivots around which national defense turned, sueltos helped loyal regional commanders maintain spatial control over large swaths of territory. They gave the federal army the "means to act energetically" at the first sign of rebellion.[113] As early as 1921 US attachés observed a sharp rise in the cavalry's flying columns and the military expediency they displayed as they traveled from Nuevo Leon to Coahuila and Hidalgo.[114] While developing sueltos, the government also planned a 3.5-million-peso cantonment in Irapuato, Guanajuato. This larger cantonment was the first of its kind—equipped with eleven barracks, officers' headquarters, a school, casino, gymnasium, and a hippodrome—and it would soon become a model to follow.[115]

Sueltos and the Irapuato cantonment helped the administration endure the formidable De la Huerta Rebellion in the winter of 1923–1924 and secure Obregón's presidential selection of Plutarco Calles. The swift completion of these defense installations helped Obregón out-engineer Adolfo de la Huerta. De la Huerta had been the nominal leader of the Agua Prieta Revolt and provisional president of the republic before Obregón was elected in 1920, but since then had been the minister of finance from 1920 to 1923. De la Huerta and the Delahuertistas that supported him wanted compensation for putting Obregón into office in the form of a formal de la Huerta presidency. Rather than revolt against Obregón, Delahuertistas opposed Plutarco Calles, the presidential candidate Obregón selected. To discredit Obregón's authority, Delahuertistas highlighted his economic relationship with the United States. By publicly denouncing the Bucareli Accords, De la Huerta openly accused Obregón of yielding unfavorably to US interests. In so doing, the presidential hopeful severed the de la Huerta–Calles-Obregón "Sonoran Triangle," and incited the most widespread military rebellion of the 1920s.

In early December 1923, when the rebels officially made their call to arms, Delahuertistas outnumbered loyal soldiers of the Ejército Nacional forty-seven thousand to thirty-five thousand. Numeric advantage notwithstanding, Delahuertistas lacked the social organization and defense infrastructure necessary to supplant Obregón's well-rooted authority. As in Obregón's ascendency in 1920, military professionalization and engineering were essential to the president's maintenance of power in 1924. The federal army defeated rebels with military-engineered cantonments, sueltos, and a nascent road network. By providing the federal army with exclusive trade rights to US arms and vehicles, Washington's official recognition widened the technology gap between the rebels. General Abelardo Rodríquez in Baja California, for instance, acquired US arms and planes like no rebel commander could.[116] To make matters worse for the rebels, the army's sueltos and Irapuato cantonment served as forwarding operating bases from which offensives incorporated an unprecedented number of automobiles and planes into their campaigns. This decisive use of planes would continue to squelch rebellions in the 1920s. From the cantonment General Amaro deployed regiments to face combatants in Guadalajara and Guerrero.[117] He ordered two hundred military trucks to fight rebels in the state of Jalisco, where they came in handy for the decisive battle of Ocotlán. In Guerrero, where Delahuertistas maintained complete state control, Amaro sent 1,500 troops down the new motor road to Iguala, where they confronted the forces of a rebel commander and regional cacique, Rómulo Figueroa. Even though Figueroa's forces managed to destroy the railroad and telegraph lines to Iguala, his battalion failed to prevent trucks from penetrating the rebel stronghold.

Obregón also gave instrumental orders from Irapuato. As the Delahuertista threat attempted to secure access to the Pacific from Oaxaca to Colima, Obregón sent weapons to the coast, secured the ports, and directed the construction of telegraphs cables.[118] With one arm, the president personally led many of his troops on the battlefield. Obregón was more successful than de la Huerta at convincing tens of thousands of peasants and small landholders to defend the nation. For example, local *agraristas* in coastal Guerrero under the command of Amadeo Vidales helped supplant Delahuertista control in Acapulco for Obregón.[119] After four months of fighting, the rebellion was over; four months later, Calles won the presidential election. The nation's military infrastructure was new and imperfect but strong enough to defeat Delahuertistas and transfer power from Obregón to Calles.

The Irapuato cantonment served as a new standard in the nationwide spread of cantonment-like bases called *campos de concentración* in 1924.[120] The bases were not concentration camps in the Spanish or German sense but were places where soldiers lived, trained, and prepared to defend the nation. Within six months the military used Amaro's cantonment model to construct concentration camps in Celaya, León, Irapuato, Guanajuato, Lagos, and Navajoa, with others in progress in Chihuahua City, San Luis Potosí, Veracruz, and Mérida. US attachés called it nothing short of a "tactical and technical reorganization of the army."[121] This ambitious cantonment craze happened only with sustained leadership. First, Obregón disbanded nineteen rebel regiments to make space in the budget.[122] Second, the army established a new engineering staff to work under the general Obregón appointed as the national commander of campos de concentración: Joaquín Amaro. The following year President Calles promoted Amaro to the secretary of war, where the general would begin to institutionalize military reforms and direct thousands of soldiers to the road-building efforts of the famed National Road Commission of 1925. Years in the construction zones of roads, encampments, and campos de concentración prepared Amaro for his formative role as the secretary of war (1925–1932).

Revolution and expedition presented Mexico and the United States with engineering challenges that reconfigured their styles of state building. The Punitive Expedition served as a training ground that helped the US military close the technological gap between themselves and European armies.[123] The expedition, like the revolution, was also a process of knowledge production.[124] Like roads, the cantonments of the US Preparedness Movement demonstrate the expedition's impact on US business and state building.[125] Highlighting how experimentation took place on the Punitive Expedition presents one way that the Mexican Revolution altered military development and state building in the United States. US military officials thought that motor vehicles could capture Villa; instead, they caught the attention of industrialists, politicians, and the public. Even though punitive experimentation did not complete its military objective, its engineers—both leaders and technicians—found new ways to build emergency roads and cantonments. Decades before President Eisenhower coined the term "military-industrial complex," Cadet Eisenhower pioneered a road network that helped establish new connections between the military and industry in the United States. Ultimately the failed expedition did not discourage military officials, politicians, and businessmen from believing that military-grade products were good business with or without catching or killing bad guys. US military officials and business leaders continued to engineer similar logic to either fund or facilitate the construction of thousands of military roads across interwar Latin America.[126]

In the Mexican case, armed conflict broke out and then reformed the once Porfirian military into a more centralized postrevolutionary form. Obregón's military reconstruction efforts, such as road building and sueltos, helped professionalize the military and foster economic development before the Bucareli Accords. This is significant because histories of postrevolutionary state building tend to begin in 1925 when the Calles administration formally institutionalized the National Road and Military Reforms Commissions. Looking at punitive engineering and defense construction from 1916 to 1924 does not refute the importance of the Bucareli Accords or the 1925 commissions but instead underscores the influence of earlier regional and military defense initiatives. Regional and military projects imbued with transnational cooperation helped professionalize the Mexican armed forces when military officials held the majority of political power. However, punitive engineering alone did not defeat the Delahuertistas, much less balance the national budget.

CHAPTER 8

FLYING MACHINES AS A MEASURE OF MEXICO

National Reconstruction, the Cultural Revolution, and the Maturation of Mexico's National Aviation Program, 1921–1945

PETE SOLAND

On June 8, 1925, Henry Woodhouse, President of the Aerial League of America, wrote to Mexican president Plutarco Elías Calles looking for his support in building the "world's first areal city" in San Marcial, Sonora.[1] Woodhouse attached an article he wrote for *Scientific Age* that laid out his plan in detail. Quoting famed explorer Admiral Robert E. Peary—known best for his Arctic expeditions—as saying such a city would be "as important to posterity as the founding of the first seaport, or the founding of Paris or London," he then explained that San Marcial's geography and topography made it ideal for such an audacious endeavor. The town was nestled close to the US–Mexican border, the Arizona–Mexico City airway, important railroad lines, several ports on the Sea of Cortez, and the Matape River, giving it immense potential as a transportation hub for shipping the region's considerable mineral wealth to markets across the globe, or so Woodhouse argued. He further projected that the capital generated by international trade might entice investors to finance the construction of a central power plant. Such a plant would in turn transform the modest pueblo into a "superpower city" with its own "superpower broadcasting station," supplying "the increasing market for radio sets in Mexico and Latin America." Woodhouse was not the only one who dreamed grand visions for San Marcial. Financiers from around the world had already taken an interest. In his letter Woodhouse sought Calles's opinion on the matter and, one might infer, his blessing.

Woodhouse's optimism for San Marcial's development proved un-founded and his vision for the town never materialized. His article never-theless reflects a sense of enthusiasm shared widely among business interests and government officials who believed that aviation held transformative possibilities as a means of capitalizing on the country's immense natural re-sources. They reasoned that aeronautics technology, if properly developed, would connect Mexico to global markets and, perhaps better still, jump-start industrial development. The siren call of globalization nevertheless posed a complicated question for politicians. Considering that the country was emerging from a decade of violent political and social upheaval brought on, at least in part, by the abuses of foreign powers exercising undue in-fluence in national affairs, and that it was currently immersed in a largely inward-looking cultural revolution, how could leaders square the cosmo-politanism of the Second Industrial Revolution with the nationalism of the Mexican Revolution?[2] Aviation seemed to offer one solution. If men like Calles could establish a viable national aviation program, then the country could participate in global affairs on more even terms.

As people worldwide adapted to a tectonic shift in technological in-novation and global interconnectivity from the late nineteenth century through the early twentieth century, a period sometimes referred to as the Second Industrial Revolution, government leaders in industrialized and industrializing nations embraced those innovations as proof of an inevi-table, mostly linear march of human progress toward a utopian future. In Mexico, as in the United States, the Soviet Union, parts of Latin America, and generally wherever nations pursued industrial modernity, state builders fixated on technologies big and small, from aeronautics and automobiles to radios and refrigerators, as crucial to national modernization efforts. Aviation, because it seemed so miraculous, provided especially compelling evidence of this technologically deterministic worldview, which was root-ed in nineteenth-century positivism and which powerful people worldwide accepted uncritically.

Policy makers took a special interest in aeronautics technology as it related to modernization and state building, but Mexicans from all walks of life participated directly and indirectly in national conversations about the technology. Mass communication allowed news of aviation to be dis-seminated to larger, far-flung audiences, and although it is difficult to gauge popular interest in aviation in Mexico from the 1920s through the 1940s, the steady increase in coverage of aviation-related news during these decades suggests that audiences with access to radios, newspapers, and magazines found the topic compelling and presumably relevant to their lives. Semana Aerea (Air Week) celebrations routinely made the front page of national newspapers and filmmakers captured the events for newsreels. Spectators and media members gathered to watch when celebrity aviators embarked on and touched down from record-breaking flights. Some pilots made a living, or at least supplemented their income, traveling with itinerate air shows to smaller, more provincial towns. When they did, they drew crowds from the surrounding areas, which were often rural.[3] In Mexico, as in the United States and Western European nations, aviation provided popular audiences with entertainment that was unquestionably modern and usually patriotic.

Popular engagement with aviation did not stop with spectatorship. Regional and local aviation "clubs" shouldered much of the burden of establishing a national aviation network, especially early on.[4] These groups campaigned to raise popular support and funds to help establish aviation in their communities. People, especially young people, from across the country were invited to consider a career related to aviation, although the path for most was difficult at best. That said, not everyone welcomed aviation projects. The construction of airports, airstrips, radio towers, meteorological equipment, and other infrastructure brought grave concerns about safety and land use to urban and rural communities alike, and community members took action to protect their best interests.

Scholarly debates about technological determinism loom large over the historiography of technology. Historians have responded by embracing theories that help them balance technologies' often profound impact on history with the importance of recognizing human agency. Thomas Hughes's concept of technological momentum, which asserts that external factors like political, cultural, and social movements more easily and directly influence a nascent technology than one that people already accept as an established part of their society, helps us better understand aviation history in Mexico from the 1920s and 1930s, a period that aligns with government-led efforts at national reconstruction and modernization following the most destructive phase of the Mexican revolution (1910–1920).[5] As Mexicans everywhere worked to physically rebuild the country and make sense of the destruction and bloodshed of the revolution, federal officials and intellectuals viewed technological advancement as essential to reimagining national identity and recasting Mexico's position in the world.

Aviation developed within the context of larger, government-led projects aimed at reengineering Mexico both physically and culturally during the period of revolutionary reconstruction through World War II (1921–1945). The cultural initiatives of the period recast national identity as the product of a mythologized past that glorified pre-Hispanic societies while dismissing modern native people. Government officials, media members, and business interests imprinted the nascent aeronautics industry with an aesthetic that reflected this vision of the nation and affirmed the regime's commitment to revolutionary nationalism.

Scholars would be wise to view critically attempts by officials to dress aviation in the clothes of a social revolution. Aviation's transformation from a relative novelty in the 1920s and 1930s to a crucial part of the national interest in the 1940s aligned closely with a shift in national politics. Manuel Ávila Camacho's administration (1940–1946) pivoted away from Cárdenas-era attempts to balance industrial modernization with social justice, and once again fixated on machines as the primary measure of Mexico's advancement. Mexico's wartime alliance with the United States and the rise of tourism as a major economic sector gave aviation the momentum it needed to become an indispensable part of national society, but the expansion of the tourist-centric commercial airline industry empowered wealthy, often foreign interests while undermining the social reforms pursued by the previous administration.[6] This chapter argues that aviation camouflaged contradictions between the revolution's more radical cultural and social goals and the more liberal development strategies that endured after the fall of the Porfirio Díaz regime.

AVIATION, CULTURE, AND REVOLUTION

Many revolutionary leaders saw aviation as a new solution to an old problem that had plagued central governments since at least the colonial period. Federal officials in the 1920s and 1930s hoped to use airplanes to unite a country where political regionalism, exacerbated by Mexico's size, diversity, and rugged geography, prevented full and effective national unification. In reality, while human flight provided a novel and convincing spectacle of industrial progress, using technology to physically connect Mexico's periphery to its center was not as straightforward as advertised.

In the latter half of the nineteenth century the railway embodied the shared desires of Mexico's ruling class to see their country modernize rapidly. They believed that attaining a fixed, positivistic modernity would act as a panacea for the country's postindependence ills. The association of railroads with ideas about progress and modernity grew during the Porfiriato, the period in which Díaz dominated national political life (1876–1911). Those of the "Porfirian persuasion," especially *científicos*, imbued trains with great symbolic importance, which they codified into law and disseminated in the press.[7] Porfirian officials dispensed with the legal requirement for a formal trial in cases of people caught interfering with railways or telegraph lines. Summary punishments permissible under the law included execution. The government-subsidized literary publication *La Revista Moderna*, founded by científico Rafael Reyes Spíndola, expounded regularly on the topic of railroads as a reoccurring theme.[8] A strong ideological continuity linked the Porfirian fascination with locomotives and the revolutionaries' decision to embrace aviation as a symbol of modernity. Leaders in both eras placed their faith in industrial progress and propagated a discourse that conflated mechanical innovation with national prestige and international importance.

Francisco I. Madero's uprising against Díaz's government in 1910 ushered in the most dramatic national transformation since independence and marked the first major social revolution of the twentieth century. Madero's call for "effective suffrage, no reelection," which mimicked Díaz's slogan when he first took control of the federal government in a *golpe de estado* in 1876, posed a narrowly framed, politically oriented challenge to the country: overthrow a corrupt dictatorship and institute a more democratic system. The widespread, populist movement that took shape in the wake of that challenge led to demands for social reforms to benefit the working class, especially peasants. These demands were embodied in Emiliano Zapata's 1911 Plan de Ayala, which made land reform a central issue of the revolution. When Victoriano Huerta plunged the country further into internecine fighting by betraying Madero and overthrowing his government, he also unleashed a greater degree of political chaos. Revolutionaries across the country broke off into armies that varied greatly in terms of size, cohesion, and political ideology.

The political situation in Mexico during the first decade of the revolution was complex, with regional contexts playing an important role, but certain divisions eventually emerged. Venustiano Carranza's Constitutionalists focused on the need for democratization but were divided about the importance of social reforms. Their leadership appeared, and often was, urbane and middle class. In opposition, the respective armies of Francisco "Pancho" Villa and Zapata were more rural and decentralist. They pushed for social reforms. Zapata especially saw agricultural workers as the revolution's principal drivers and beneficiaries. Neither representatives from Zapata nor Villa were invited to the Constitutional Convention in Queretaro, which Carranza sanctioned, in part, to validate his government. The convention nonetheless exhibited the divisions among Constitutionalist political actors. Intellectual clashes between the "Jacobins" led by Francisco J. Múgica and moderate members such as engineer Félix Palvancini, who advocated for more classically liberal reforms, divided the delegates. The debates, and ultimately the constitution that they wrote, illustrated the inescapable importance of social issues, especially land reform. Too many people had fought and died because of agrarian demands. National debates about land reform and workers' rights remained important themes in Mexican politics for decades, before reaching a crescendo during the Lázaro Cárdenas administration.[9]

Issues related to industrialization and technological development appeared in the national conversation during the first ten years of the revolution, even if most Mexicans viewed them as less important. The majority of people still lived in the countryside, where the effects of the Second Industrial Revolution were less obvious. Díaz's regime prioritized bringing progress to cities, mechanizing large-scale agricultural operations, and supporting industrial manufacturing, which was mostly in the northern and central parts of the country. Consequently, technological change, and problems related to urbanization and industrialization, was most visible in and near these places. In the decade leading up to the revolution, wealthier urbanites could purchase electric appliances, although few homes were wired with electricity. The first automobile arrived in Mexico City in 1895, although there were few roads suitable for travel. Radio brought mass communication to the country, but the equipment was expensive and problematic, and static was constant.[10]

Revolutionary leaders on all sides embraced new technologies when beneficial. In 1912 Madero became the first head of state to fly in an airplane. The first instance of aerial-naval combat occurred off the coast of Topolobampo, Sinaloa, in the sea of Córtez, when, under orders from Álvaro Obregón, pilot Gustavo Salinas Camiña and mechanic Teodoro Madariaga bombarded a Huertista warship from a biplane. Villa adroitly made the most of embedded camera crews to gain publicity and support for his cause and later dodged aerial observation by the Pershing Punitive Expedition. The delegates who wrote the 1917 Constitution anticipated the growing importance of industry in the country's future by including provisions such as Article 123, which "granted rights and privileges only partially expressed in the legislation of highly industrialized nations, rights that organized labor had been striving to obtain since the advent of the Industrial Revolution."[11]

The violence of the revolution slowly abated during the 1920s, allowing the government to pursue reconstruction projects aimed at fostering national unification, both physically and culturally. As Mary K. Vaughan and Stephen Lewis observed, "utopian visions surged in the heat and hope of the revolution."[12] Government officials, together with engineers, inventors, and other technical experts, took inspiration from the major cultural movements of the period such as *indigenismo* and *mestizaje*. These concepts envisioned a more homogenized national culture while promoting racist ideas that a distinctly Mexican revolutionary modernity could be achieved by fusing Europeans' supposed innate capacity for innovation with mythologized accounts of ancient native spirituality. Consequently, new technologies and industrial projects were frequently adorned with related cultural symbols and promoted through cultural events.

Cultural approaches to national modernization were on display at the ceremonies held in 1921 by Álvaro Obregón's administration (1920–1924) to commemorate the centennial anniversary of Mexico's independence. The monthlong festivities showcased the country's art, crafts, food, music, and regional dances.[13] Celebrations also provided an excuse to showcase the nation's technological achievements. New technologies proved especially effective at capturing the attention of audiences and pushing them to envision an exciting, more unified future. At Balbuena field, just outside of Mexico City, the Mexican Air Force (FAM) treated government officials to an air show that demonstrated advances in aviation and radio, another developing technology that leaders used to shape national culture and modernize communications across the country. Filmmakers, like Obregón propagandist Vicente Cortés Sotelo, used them to project an image of modernity to both national and international audiences.[14] The celebrations thus demonstrated both sides of officials' reconstruction efforts: a vision of the nation as traditional that they presented to domestic audiences, juxtaposed with a plan for modernization that they hoped would make the country more competitive globally and would be appealing to foreign investors.

Aviation's close association with cosmopolitanism in the popular imagination dated back to its invention. The rivalry between the Wright brothers and Alberto Santos-Dumont attracted an international community of inventors, pilots, and enthusiasts, whose diverse contributions to the development of aeronautic technology highlighted its globalizing potential. The technology's globalizing appeal also appeared in Woodhouse's 1925 article that promoted the construction of an air city not simply as a national accomplishment, but as a contribution to human history and global trade. In 1928 Mexico celebrated Semana Aerea, which invited pilots, engineers, mechanics, and business people from around the world to visit Mexico City to participate in and spectate events and competitions showcasing aviation. Newspapers in the capital covered the celebrations in depth and compared Mexico's accomplishments with those of Western European nations and the United States. A report from the Mexican Aeronautics Association lauded the English as a model for "air mindedness" and applauded their use of aviation to maintain the global reach of the British empire.[15] Nor were Mexicans alone in viewing achievements in aviation through racial and cultural lenses. Willie Hiatt demonstrated that in Peru aviators and aviation were hailed as exemplars of an "Andean" modernity that built off Peruvian interpretations of indigenismo and mestizaje.[16]

Although aviation's utility as a globalizing technology was, to a great extent, precisely what made it so desirable to state builders, authorities endeavored to balance the technology's worldly qualities by imbuing it with a distinctly national character. The army's Escuela Nacional de Aviación published a magazine whose title, *Tohitli*, referred to a figure in an ancient Aztec religious story who transformed into a bird to escape a world-destroying firestorm. The publication ran from 1915 to 1939 and included both domes-

tic and international aviation news. It also featured a column that reprint-
ed stories and poems with allusions to flight and Mexican patrimony.[17] By
pairing images and allusions to the country's mythologized Indigenous past,
officials neutralized, or at least obscured, the foreign influences intrinsic to
the technology. In so doing, they strengthened revolutionary patriotism and
national identity while still measuring progress with the same ruler used
by the industrial West, which had been among the most vocal critics of the
Porfirian regime before the revolution.

Cultural revolutionaries promoting secularism perceived science and
rationality as an antidote to the regressive effects of the Catholic Church
in national affairs, and crucial to modernization. The explicitly stated purpose of what critics decried as "socialist education" was to replace religious
teachings with explanations rooted in scientific thought. The Ministry of
Education under both Obregón and Calles highlighted technological innovations in lesson plans and created secular holidays that celebrated inventors in the way that religious festivals celebrated saints. Famed anticlerical
governor of Tabasco Garrido Canabal elected to fly in an airplane painted
red and black, colors synonymous with his radical anti-Church agenda, and
Tabasco newspapers printed poems that invoked the airplane as imagery
demonstrating the power of science in hopes of goading the clergy.[18]

Secularizing reforms in Mexico during the 1920s and 1930s rarely de-
served the "socialist" label adopted by critics, but the approach of using
of aviation to counter the influence of the Roman Catholic Church resembled the Soviet government's use of aviation to undermine the Eastern
Orthodox Church. During the 1920s the Red Air Fleet "baptized" villages
with flyovers, and the Soviet government replaced religious iconography
with depictions of pilots and airplanes.[19] Similarities in how Mexico and
the Soviet Union used aviation as a cultural tool to facilitate secularization reveal Mexico's unique position in global affairs and highlight the two
countries' unique relationship. Despite Mexico's border with the United
States, and the fact that Mexico's aviation community viewed the United
States and Western Europe as aspirational models, Mexico and the Soviet
Union exchanged technology and culture while cooperating in global affairs throughout the period.[20]

AVIATION UNDER MILITARY STEWARDSHIP, 1920–1928

The military, whose officers saw themselves as the caretakers of the revolution, shaped aviation culture as much as or more than any other group in
Mexico. It was one of the only institutions with the material and monetary
resources to develop the technology. Most people who worked in the field
of aeronautics received their first training while serving in the military. Air
shows and goodwill flights sponsored by the military garnered considerable

attention in the press, providing officials with an opportunity to promote patriotic values and define civic virtues. Aviation's military application also appealed to political leaders wishing to bring order to the country. Presidents Obregón and Calles used the air force to counter rebellions and attack those who did not conform to the vision of national identity favored by the new regime.

Military-led state-building projects during the initial phase of the revolution and reconstruction proved immensely influential in the development of emerging technologies. Justin Castro's previous work and Jayson Porter's chapter in this volume demonstrate this in regard to radio and road-building projects, respectively. Their observations hold true for aviation. Just as with the development of radio and roads, military stewardship over aviation took place within a broader effort to professionalize and modernize the military.[21] Furthermore, Obregón and Calles both successfully used the infant air force to consolidate political control during the 1920s. Growing faith in aeronautics' military application led them to increase government investment generally. Calles began his admiration for aeronautics technology after Interim President Adolfo de la Huerta made him the secretary of war in 1920. Calles worked with US pilot Rafael O'Neill to reorganize and strengthen the Mexican Air Force by purchasing planes from Europe, because the United States had embargoed the sale of aircraft to the country after the Agua Prieta coup toppled Carranza's government. Somewhat ironically, Calles used the newly modernized air force to help defeat de la Huerta's rebellion in 1924, when de la Huerta and a significant portion of the country's military officers attempted to topple the government after Calles's election to the presidency.[22]

As governor of Sonora, Calles employed military airplanes to dispel remnants of armed groups loyal to Villa in 1918. He also used them to violently suppress an insurrection by the Yaquis of Sonora that same year, and then again as Mexico's president in 1926. Government attacks on the Yaquis demonstrated the limits of revolutionary leaders' commitment to honoring the country's Indigenous people, a crucial part of the cultural reconstruction of the nation. Those swept into power after 1910 wished to strengthen nationalism by drawing on Mexico's pre-Hispanic past, albeit a very mythologized and sanitized version of it that left little room for acknowledging modern native societies. The pressure to compete in global markets convinced officials that attaining the order and progress promised by the ancien régime was still in the country's best interest moving forward. To men like Calles, Obregón, and the pilots who flew missions during the conflict, the campaign against the Yaqui rebellion represented the triumph of modernity over backwardness and savagery. From this perspective the air force epitomized the unstoppable charge of progress. The campaign highlighted the government's conflation of technological progress with moral superiority, which officials used to justify the violent pacification and removal of a people who dared to question why national political leaders failed to enact the reforms promised to them in 1920.[23]

Early airplane manufacturing in the country began during the revolution with the army's establishment of the National Aeronautics Construction Workshop (Talleres Nacionales de Construcciones Aeronáuticas, TNCA) in 1915. Under the direction of Italian airplane mechanic Francisco Santarini and Juan Guillermo Villasana, who worked on the construction of the first airplane assembled in the country, the workshop represented both an attempt to start a viable domestic manufacturing program and a way to offset the US embargo on selling aeronautics equipment to Mexico. From 1915 to the mid-1920s the TNCA developed four series of planes, including an 80-horsepower general trainer that served as the backbone of the country's embryonic airmail service.[24]

Ángel Lascurain y Osio took the helm of the TNCA in 1920 and, together with mechanical expert and military officer Juan F. Azcárate, improved designs for building more practical, moderate-sized aircraft that were better equipped for general-purpose use. This included an endeavor to build a sturdy and efficient plane capable of navigating the short runways encountered in the country's dense jungle and mountainous terrains. The designers named their planes after national symbols. The Sonora drew its name from the home state of President Obregón and future President Calles. The Quetzalcoatl harkened back to the winged serpent of pre-Colombian religion. Last, the Tololoches made a sound reminiscent of the pre-Colombian drum.[25] The appearance of national symbols in the domestic production of aircraft represented attempts to ignite excitement about the country's future glory by summoning pride in its pre-Colombian past and its revolutionary legacy. Just as anthropologist Manuel Gamio's indigenismo built on the use of Aztec culture as a symbol of Mexican nationalism in the early twentieth century, so too did officials and intellectuals use pre-Colombian symbols to imbue technological and industrial projects with patriotism. As Marcela Saldaña Solís notes in her chapter in this volume, the Díaz government constructed a statue of the last Aztec emperor Cuauhtémoc on the Paseo de la Reforma, blending pre-Hispanic imagery with sophisticated, cosmopolitan city design. Similarly, in 1890 the brewery Cervecería Cuauhtémoc took inspiration for its name and logo from the very same emperor.[26] The emperor's name was even applied to dredges used in canal construction.[27]

In 1927 Calles appointed Azcárate the director of the TNCA. Collaborating with test pilot Louis Boyer, who later worked as Compañía Mexicana de Aviación's (CMA)—popularly known as Mexicana—director of pilots, Azcárate developed a new sesquiplane that became the crowning achievement of early domestic manufacturing. Officials adopted the Sesquiplano Azcárate as the training plane for the Military School of Aviation, and its safety record earned it fond memories from the generation of pilots that flew it. Azcárate's leadership of the TNCA marked an important turning point in the maturation of domestic aircraft manufacturing, as he attempted to advance manufacturing beyond a solely government-run endeavor by establishing a private domestic industry.

In 1930 Azcárate founded the country's first private aircraft manufacturing plant with financial investment from General Abelardo Rodríguez, under whom Azcárate later served as his chief of staff. He hired Lascurain y Osio to help with engineering. On May 29 Azcárate's company, Companía Juan F. Azcárate, entered into a contract with the Ministry of War and the navy to construct seven Fairchild- and thirty-one Corsario-model airplanes for a total of 1,533,470 pesos. Nevertheless, the continuing global depression badly hurt the company, which soon overstocked the air force with its planes. Azcárate looked south in an attempt to keep his business buoyant, hoping to sell the aircraft to Central and South American countries with similar development needs as Mexico, but with little success.[28]

The demise of Companía Juan F. Azcárate constituted a major reversal for airplane construction in the country. It also illustrated the young technology's vulnerability to external conditions, in this case the depressed economy. Major domestic manufacturing did not reemerge until after World War II. This was probably because, unlike with lighter manufactured goods, demand for heavy machinery like airplanes decreased more rapidly, could be met by established manufacturers already dominating the markets, and recovered more slowly than demand for nondurable consumer goods. Nevertheless, early attempts to establish a domestic aircraft industry had the advantage of cultivating a cadre of aeronautics experts vital to the development of the country's civil aviation program.

TRAINING AERONAUTICS EXPERTS IN RECONSTRUCTION-ERA MEXICO

National reconstruction provided officials with an opportunity to modernize and industrialize the country's economy and infrastructure, but such changes required transforming a largely rural, agrarian population into a more urban, technically savvy workforce. The Military Aviation School furthered these plans by training students for jobs as mechanics, mechanical engineers, and pilots. The education they received reflected positivist definitions of progress focused narrowly on technological growth and material wealth.

Officials saw the standardization of knowledge and training as crucial in formulating a corpus of aviation professionals comparable to those in industrialized countries. Efforts by Obregón and Calles to professionalize the military, which they prioritized to curb the frequent rebellions that threatened national stability, extended to the Military Aviation School. Acting on the suggestion of Colonel Rafael O'Neil, President Obregón decreed that all pilots, even those who had learned to fly before the establishment of the school, needed to pass a training program to confirm that they grasped modern flight protocols. The order vexed veteran pilots, many of whom felt that it cheapened their achievements and undermined their status in the eyes of younger pilots. O'Neil remained resolute despite the controversy, and eventually veteran pilots acquiesced. In his memoir pioneer aviator Roberto Fierro praised O'Neil's steadfastness. He argued that O'Neil's unwillingness to acquiesce to claims of seniority privilege ensured a higher level of technical expertise among the nation's pilots going forward. In doing so O'Neil put into practice contemporary attitudes toward professionalization and demonstrated the thinking behind a growing movement to reform the military.[29]

Federal aviation authorities endeavoring to implement standards for pilot training and service contended with the constraints of an immature aviation program that lacked the number of experts necessary to draw firm distinctions between military and civil aviation. The shortage of trained pilots meant that military aviators piloted civilian flights. In 1929 the Federal Air Service employed nine pilots who were studying at the Military Aviation School. The assignment earned the pilots an additional ten pesos a day, bringing their daily salary to twenty-three pesos.[30] The situation posed officials the challenge of remaining flexible enough to respond to their financial limitations while maintaining a degree of professionalism among military pilots. President Abelardo Rodríguez (1932–1934) attempted to solve the problem by issuing a decree that allowed military pilots to fly for private companies while simultaneously implementing regulations designed to ensure that they remained ready for military service.[31] The order typified the compromises and stopgaps that officials felt pressured to make during the program's early stages.

Revolutionary leaders took an active interest in their citizens' health and physical aptitude as part of their aspirations to modernize the country. Delegates had established this interest in Article 73 of the 1917 Constitution by empowering the federal government to monitor and protect the general health of the country.[32] Aviators, as archetypes of authorities' vision for national progress, served as logical targets for government oversight of the human body. Pilots underwent testing and training to evaluate their mental acuity, technical knowledge, and physical fitness, all of which officials deemed important indicators of an aviator's competence.

At the country's first civil aviation school, the Escuela de Aviación Civil "Emilio Carranza," students underwent thorough medical examinations. These exams were filed with the Department of Civil Aeronautics, along with monthly reports that included their personal information, flight hours, and any accidents in which they were involved. The exam featured large portions devoted to testing a pilot's vision and nervous system, unsurprising concerns given their importance to flying an aircraft. Physicians also recorded information about what sports students played, their consumption of tobacco and alcohol, and their sex drive.[33] Beginning in 1930 the department kept files on mechanics who obtained a special license to work on aircraft. Aircraft mechanics often served on flight crews in case mechanical problems presented while in the air. Like pilots, mechanics' department files included, in addition to personal background, the results of a medical exam, along with an evaluation of their theoretical and practical knowledge of aircraft mechanics.[34]

Establishing a baseline for qualifications of aviation experts allowed the government to clear an important hurdle blocking the development of a national aviation program. Their efforts to standardize knowledge of aeronautics and to professionalize practitioners fit with modernizing discourses espoused by the progressive reformers of the day.[35] Government sponsorship of aviation during the early and mid-1920s followed a general strategy whereby officials used direct and indirect investment in transportation infrastructure and construction to help nurture the development of domestic industry.[36] By 1928 these policies produced a pool of experts who oversaw the next stage of aviation growth. This group shared a common set of experiences based on military service and formed a tight-knit community of nationalist yet cosmopolitan professionals who supported investment in technological and industrial development.

THE DEPARTMENT OF CIVIL AVIATION

On December 11, 1929, President Emilio Portes Gil (1928–1930) stood inside the main hangar of the recently constructed Central Airport at Balbuena Field to inaugurate the Great Exposition of the Aeronautics Industry and officially kick off national Semana Aerea celebrations.[37] Upon finishing his speech to an audience of diplomats and business leaders from around the world, the president boarded a new model six Ford Trimotor airplane owned by the country's largest airline, Mexicana. Two of his top-level cabinet officials, as well as delegates from England, Italy, and France, and CMA's vice president Espinosa Mireles, followed the president on board. The plane took off in front of an excited crowd of onlookers. Once out of the city, the pilot circled the Valley of Mexico, treating the passengers to a view of the famous Popocatepetl and Iztaccíhuatl volcanoes, two of the nation's oldest landmarks and symbols of its grandiosity.[38] The flight encapsulated the central goals of the week's celebration: to showcase the strides that officials had made since Plutarco Elías Calles fast-tracked the national aviation program the prior year, and to seat aeronautics industry leaders next to potential foreign investors.

In 1928 President Calles authorized the creation of the Department of Civil Aeronautics, signaling the government's intention to take seriously the technology's development as a tool for economic expansion and national modernization. The Ministry of Communications and Public Works (SCOP) had previously housed the Technical Section of Air Navigation as a subsection of the Department of the Railroad, a testament to the marginal status of aviation in government plans at that time. Officials granted the first concession to provide regular passenger, mail, and cargo service in 1921 to the Compañía Mexicana de Transportación Aérea to provide service from the capital to cities along the country's central and northern Gulf regions. In 1924 the company merged with Compañía Mexicana de Aviación, quickly becoming the largest airline in the country.[39] Civil air routes nevertheless developed slowly throughout the decade. The creation of the Department of Civil Aeronautics was intended to speed up the expansion of the air network through increased funding and better administration.

Tasked with overseeing all civil aviation in the country, public and private, Department of Civil Aeronautics officials' primary concerns included streamlining air postal service and improving infrastructure by building airports and airstrips. Under the direction of the veteran airplane designer and former head of the Military Aviation School, Juan Guillermo Villasana López, the new department had a mandate to purchase more planes and equipment, and the department's budget included funding for the construction of a central airport in the capital.[40] Officials planned to create an air postal network that functioned as an umbrella, connecting the major state capitals with one another and linking them all to Mexico City. President Calles authorized Villasana to hire personnel, with the explicit understanding that he and future heads would privilege applicants from the military and take on civilian pilots only as necessary. The policy established ideals of national service as central to the ethos of the department from the time of its foundation.

Problems with official plans soon appeared: the department lacked the capability to effectively provide service to such a vast area, despite having received considerable support from the president. The executive office hoped to extend the reach of air post by granting the ability to negotiate contracts with private companies, much as the government had used similar arrangements to expand the national highway system.[41] How Callista officials structured and implemented civil aviation reform closely resembled their earlier modernizing initiative, revealing an underlying pattern in the administration's strategy for achieving national progress. That is, in both cases they used direct and indirect investment in transportation infrastructure to better integrate the provinces into commercial markets and foster a sense of national connectivity.

AVIATION AND INFRASTRUCTURAL EXPANSION DURING NATIONAL RECONSTRUCTION

Calles's efforts to improve infrastructure are well known, especially his focus on railroads and highways.[42] Porter's chapter in this volume highlights how Calles's reputation as a state builder overshadowed Obregón's earlier infrastructural initiatives and perhaps obscured an early, but crucial, part of revolutionary reconstruction. This point holds for aviation as well. Comparing the government's roads program with its civil aviation program further enriches the story of Mexico's modernization during this period, allowing some generalizations about how such programs operated and the effect of reforms on the emerging industrial sector. The common logic behind both Obregón's and Calles's modernizing reforms and their reliance on clientelist arrangements—that is, a system of political patronage that centralized power by governing through an exchange of personal favors—shaped how authorities managed aviation well into the 1930s. The similarities illustrate the degree to which governance, reconstruction, and technology were mutually constructive during the era. For both Obregón and Calles, making progress toward the industrial development of the nation was an important part of creating stability and maintaining political control.

The National Road Commission's extension of the country's highway system and the Department of Civil Aeronautics' establishment of an air network were attempting to solve problems posed by the country's geography. As indicated in the department's founding documents, officials prioritized establishing air routes that connected the country's major cities and facilitated international exchange.[43] The policy centralized oversight in Mexico City before channeling it to state capitals, where it could be further disseminated to local communities. The network bolstered the federal government's ability to deliver mail, a crucial step toward addressing the problems that had caused mail service to decline both in terms of total routes and total distance covered by those routes from 1918 to 1927. The hardest-hit places included rural areas, especially where decades of fighting had damaged infrastructure. Instability caused by the revolution, especially in the north, led to the creation of the *sección postal militar*, which paired military guards with mail couriers to ensure delivery.[44] Reestablishing a functioning postal service throughout the country was necessary for asserting central government control, and building roads and airstrips was the best way to improve mail service over the country's vast and often rugged terrain.

The creation of a civil aviation network, like the roads project, relied on a mixture of government, military, and private initiatives. Private companies helped pay for and constructed the national aviation infrastructure, although they often benefited from government subsidies and military assistance.[45] By 1927 the government had granted four air concessions to private companies, but service remained limited to the central and northern Gulf regions. By January 1934 sixteen companies operated concessions spanning approximately 14,500 kilometers that covered much of the country. Workers for the national roads program had, by contrast, constructed close to 10,000 kilometers of roads between 1925 and 1940 and had made only limited headway into the southern and northern parts of the country. The relatively fast pace with which aviation expanded highlighted one reason it appealed to so many politicians and industrialists: basic aviation infrastructure required little more than the construction of an airfield, equipped with a radio tower, airplanes, and pilots. While road travel remained a more common feature of everyday life for most people, the time and labor required to expand the nation's highways capped the rate at which the program could progress. As Henry Woodhouse noted when he quoted the pioneering meteorologist Abbott Lawrence Rotch, "The air makes every place an aerial port, and affords the shortest distance to any other place."[46] Airfields functioned more like a chain of islands connected by flight paths, giving developers a way to bypass considerable environmental obstacles to connect even remote locations to national life.[47]

The Department of Civil Aeronautics was similar to the National Roads Commission in that both instituted a nationalist program promoting industrial modernization and expanding the domestic workforce. Preferential hiring for military pilots rewarded their service to the *patria* and kept them employed in the government. The laws governing civil aviation reinforced the department's nationalist mission, protecting the interests of the domestic sector with robust regulations over the industry. As part of the 1930 Civil Aeronautics Act, foreigners wishing to obtain concessions to deliver mail, people, or cargo needed to incorporate their business as a Mexican company. The act also required all companies to have at least one Mexican citizen sitting on their board of directors and to employ a workforce comprised of 80 percent Mexican citizens overall and 33 percent of management and administrative staff. Additionally, it required companies to deposit money in the country's central bank, the Banco Nacional de México, another institution created as part of Calles's reformist agenda.[48] The 1932 General Communications Law expanded protections, requiring all pilots for domestic flights and two-thirds of the rest of the crew to have been born in the country. Authorities used these measures to promote job growth among citizens and prevent foreign interests from exercising too much influence in the aviation industry.[49] The protectionist policies of the Department of Civil Aeronautics reflected the impulse to develop industry and assert national control over the economy that was characteristic of many of the Callista reforms.[50]

Calles created the National Road Commission in 1925, much as he would the Department of Civil Aeronautics three years later, to foster economic growth by facilitating transportation and trade and to pave the way for large-scale industrialization. Officials also saw both the road and aviation networks as essential to reforming citizenship by facilitating engagement in national life and teaching the public to behave as modern consumers. Both transportation projects further presented an opportunity to employ large numbers of Mexicans. In both cases the administration faced a lack of homegrown professionals, so the government eventually granted contracts to foreign companies. In both programs officials included stipulations to ensure that the majority of their workforces remained nationals to offset this concession to foreign interests.

CARDENISMO AND NATIONAL AVIATION

The Cárdenas administration (1934–1940) left an indelible mark on national aviation. The industry regained momentum after suffering stagnation in the early 1930s, as the economy recovered from the Depression. Cardenista officials took advantage of these improved conditions to reorganize and expand civil aeronautics, transforming it to reflect their attitudes toward industry. They desired to both reduce foreign influences while increasing their access to the technology, training, and materials needed to expand the nation's air network. Authorities grew ever more preoccupied with the latter concern over the course of his term, as the domestic and international political climate appeared increasingly unstable toward the end of the decade. The intersection between the politics of the Cárdenas era and aviation development illuminates the shifting priorities of the national party as it sought to industrialize and expand its presence in global affairs while maintaining a commitment to social justice. These goals ultimately required sacrifices, particularly from the countryside, as policy makers shifted their focus away from land reform at the end of the decade.

Cárdenas's presidency marked a sharp change in the country's political direction. The new president made social justice a cornerstone both of his campaign and the focus of his administration's most publicized initiatives when he first took office. His ambitious agenda soon put him at odds with Calles, whom many contemporary observers presumed to be the country's unofficial *jefe maximo*. The two clashed over issues such as land reform and the nationalization of the country's oil industry. Cárdenas eventually won the political battle, and Calles left the country in exile.

Cárdenas saw industrialization, including the mechanization of agriculture, as essential modernization and compatible with his administration's focus on social justice and for Mexico to hold its own in global affairs.
The nation's industrial and landed elites were nevertheless concerned with
the president's ejido program, which they viewed as radical for redistrib-
uting land to peasants and other poor agricultural workers. Critics of the
Cárdenas administration also balked at what they condemned as reckless,
socialist policies that too frequently favored labor unions and national interests over more business-friendly free trade policies. [51] A closer look at how
the country's national aviation program changed during Cárdenas's administration presents a more complex picture. Certainly some Cardenista-era
policy decisions were strongly nationalist and sought to empower the working class, but his administration also did a great deal to advance technological development in the country, including aviation, although financial
constraints continued to limit the speed at which this happened. In 1938
Cárdenas moderated the aggressiveness with which his administration pursued land reform and increased government support of industrialization,
most famously by announcing his intention to expropriate the oil industry.
He also reorganized the Partido Nacional Revolucionario (PNR) into the
Partido Revolucionario Mexicana (PRM) that same year, in anticipation
of the growing backlash from an alliance of conservative forces, led by industrialists and former Callistas. These decisions helped pave the way for
the 1940 election, during which Manuel Ávila Camacho won his party's
nomination over Francisco Múgica, and which ushered in a period of rapid
industrialization. [52]

In 1938 Elfego Castaneda of Tehuacán, Puebla, wrote to the president
pleading for assistance in realizing his lifelong ambition to become a pilot.
The eighteen-year-old Castaneda laid out his case in effusive prose. He explained that Cárdenas's beneficence in helping thousands of others inspired
him to ask for help. His family, although poor, held deeply patriotic beliefs,
and his brothers planned on serving in the military when they grew old
enough. He already possessed the necessary determination and needed only
some financial support from the government to begin his studies. [53] Castaneda's letter was one of hundreds of such pleas that the president's office
received during Cárdenas's term. [54]

The case of Miguel Carrillo Ayala, who, with no formal training, built
a functioning airplane mostly from scratch and flew it from Zitácuro, Michoacán, to the military base at Balbuena Field in Mexico City, provides
a stunning example of everyday people's capacity for mechanical ingenu-
ity and enthusiasm for aviation. Born in the small village of Agostitlán,
Michoacán, Carillo Ayala learned about aeronautics from reading essays
written by famed pilot Roberto Fierro and the magazine *Popular Mechanic*.

As a young man he moved to Zitácuro, where he obtained a 1930 Ford four-cylinder engine and accompanying radiator. With the help of master mechanic José Zepeda, the duo refashioned the engine by stripping away some parts, increasing its horsepower from 10 to 35, overhauling the cooling system, and replacing the distributor with a magnet. He relied on donations from friends, family, and aviation enthusiasts to purchase a wooden propeller from Mexico City and, together with a pair of brothers who were his childhood friends and grew up to be carpenters, constructed the fuselage. Carrillo Ayala dubbed the plane *Pinocho* and after a number of test flights in 1935 and 1936, he made plans to undertake the historic flight to México City, which he did, successfully, on May 14, 1936. The heroic feat made front-page national news and remains a proud part of Michoacán's technological patrimony.[55]

Requests like Castaneda's and the folkloric-esque story of *Pinocho* illustrate both aviation's popular appeal and the obstacles that impeded everyday people from pursuing a career in the field. At least some poor and rural people saw a career in aviation as a path to advancement, although it is impossible to know how many, and class structures made it difficult for those interested in training. Aspiring mechanics, engineers, and pilots needed to obtain a secondary school education before administrators would even consider admitting them to their desired program. They needed money to pay tuition and enough financial stability to allow them to concentrate on their studies instead of working full-time. On top of this, few such programs existed in the 1930s. Those who did not wish to join the military had fewer options still. Most had to uproot their lives, leaving their communities and traveling long distances to attend an air school. These obstacles limited social equality and preserved the image that the air remained the dominion of a privileged few who enjoyed the means to spend years training at one of a handful of institutions. This, almost categorically, excluded the rural poor.

Despite these shortcomings, Cárdenista officials did succeed in upgrading the country's ability to train aeronautics experts by the late 1930s. The enthusiasm with which everyday people from all over the country embraced aviation, not as an abstract symbol of the country's modernity but as a career, nevertheless proved too much to accommodate. The dissonance between the Cárdenas government's desire to promote social equality in the industry and what it could reasonably do to cultivate expertise demonstrated how material restraints continued to hamper national growth after the end of the Depression. This started to change in the final years of the president's term, with the opening of the Instituto Politécnico Nacional's aeronautics-engineering program and officials' increased efforts to find funding for schools and students. Although these additions did not amount to a sea change in the country's production of pilots, engineers, and mechanics, they nevertheless mirrored an enduring shift from agriculture to industry as the focus of the economy that, along with the onset of World War II, led to a dramatic expansion in aeronautics training.

AVIATION AND THE EL FRESNO EJIDO

As the national aviation network expanded, government officials and the private interests backing the industry demanded more land in order to facilitate continued growth. Their needs conflicted with the Cárdenas administration's emphasis on providing support to the rural poor. Cárdenas's first major undertaking as president had been to embark on a historic land reform policy that included the most ambitious land redistribution program since the start of the revolution. His supporters hailed the program as the closest any president had come to realizing the promises made in the Constitution of 1917. Nevertheless, after 1938 the federal government increasingly focused efforts on nurturing the industrial sector, causing critics to claim that the president had abandoned agrarian communities.[56] Instances where politicians and industrialists clashed with campesinos over ejido lands offer striking examples of the upheaval caused by the country's transformation into an industrial nation and a shift away from the focus on social equality in the countryside that had defined the early years of Cárdenas's presidency. One such example, involving the El Fresno ejido, provides a particularly striking example of how the process unfolded.

The expropriation of land belonging to the El Fresno ejido of the La Comarca Laguna region highlighted the severe power imbalance that dictated interactions between airlines and rural communities. The dispute was especially poignant considering the symbolic importance of La Comarca to the Cárdenista land reform program, and because it exemplified what historian Adolfo Gilly and others meant when referring to an interruption of the revolution.[57] Large-scale cotton and wheat cultivation developed on the heels of irrigation projects during the mid- and late nineteenth century. The area became a model for modern agriculture during the Porfiriato, but inequality and poor labor conditions among agricultural workers and peasants pervaded and persisted after the revolution's onset. Frustrations led to a broad-based labor movement that culminated in a massive strike beginning on August 19, 1936. The strike ended on September 3 when Cárdenas redistributed the land possessed by large haciendas to the workers. The president patterned his land reform program after the La Comarca Laguna redistribution. The region that once epitomized the Porfirian vision of agricultural modernization now exemplified the Cárdenista model, which used the achievement of social equality as a partial measure of national progress. The government's endorsement of this model nevertheless proved short-lived.[58] In 1944 the Ávila Camacho government handed over some of the lands to the airline Companía Lineas Aéreas Mineras to build a new airport for the city of Torreón, the region's center.

For *ejiditarios* like those of El Fresno, the progress that commercial airlines and government officials sought came at the expense of what they had gained from land reform. The loss of nearly half their territory spelled disaster for the El Fresno ejido. About a year after the press printed the decree, leaders writing to the president from the Hotel Merida in Torréon complained that they were still displaced. The airline had built on the productive land, leaving behind the territory that the state government's commission had described as too salinized to be arable.[59] Their experience highlights the complexities and costs of aviation, and industrial modernization in general, which many national and regional officials were quick to gloss over or dismiss.

WORLD WAR II, TOURISM, AND ÁVILACAMACHISMO

Aviation in Mexico underwent profound changes during the 1940s. The selection of Manuel Ávila Camacho as president (1940–1946) cemented a shift toward industrialization as the primary focus of the government's economic development programs. This shift greatly benefited aviation generally and commercial airlines specifically. Although Cárdenas had laid the groundwork for emphasizing industrialization over agriculture when he reorganized the ruling party into the PRM in 1938, Ávila Camacho's markedly warmer approach to foreign and domestic business permanently pivoted the country away from the socially oriented agenda of the previous administration. Policy makers instead looked to market growth and technological development as the primary indicators of national progress. The new president's agenda, combined with the wartime alliance with the United States and an increase in tourism, provided aeronautics technology and the commercial airline industry with crucial momentum that pushed them past the obstacles of the previous decades. Civil aviation came out of the war sufficiently interwoven with society to render it inseparable from the national interest.[60]

Ávila Camacho's administration played a crucial role in instigating the transformation of aviation in the 1940s. The president signaled his administration's commitment to improving the nation's civil air capabilities in his 1943 Message to the Nation. In it he briefly but emphatically made the case for government investment in civil aviation, arguing that the country could expect to reap social and economic benefits that would dwarf the costs of developing aeronautic and related technologies.[61] He also used wartime powers to take a more active role in directing the economy and curtailing labor movements, the consequences of which helped move commercial aviation from a peripheral industry to a central part of the government's new economic strategy. His administration cooperated with the US government, Pan Am, and its subsidiary, CMA, to push out competition that was alleged to have cooperated with the Axis Powers.[62] He also heeded industrialists' calls to reign in labor, using the state of emergency caused by the war to secure promises from unions that they would not strike. Carpenters and workmen in Mérida, Yucatán, complained that the Compañía Aeronautica Francisco Sarabia abused this agreement during a contract dispute in June 1942.[63]

US wartime aid provided the material support necessary to update Mexico's air capabilities, making the expansion of commercial aviation possible. Although the most famous outcome of Mexico's alliance with the United States was the creation of the Mexican Expeditionary Air Force squadron known as Escuadrón 201, the partnership between the two nations had a more lasting effect on the civil sector. Officials' cooperation with the United States secured Mexico access to equipment and training that aviation professionals had sorely lacked before the 1940s. The US military provided two dozen aircraft between March and September 1942, partially through the Lend-Lease program. Civil aviation schools, many of which chronically lacked the necessary planes and equipment to properly train students, received these as hand-me-downs from the government.[64] US aid also included training for pilots, mechanics, and engineers. The Ávila Camacho and Roosevelt governments collaborated to create two civil air schools, one in San Antonio, Texas, and the other in Ciudad Puebla, Puebla. The United States agreed to provide ten light aircraft to each school and eight scholarships to Mexican students. A highly successful US Civilian Pilot Training Program augmented Mexican military training programs, which contributed to the expanded pool of aviation professionals available after the war. This included members of Squadron 201, who entered the commercial sector upon leaving the air force, as well as three pilots who later flew for presidents.[65]

The greatest impetus for the expansion of commercial aviation came from a significant increase in tourism during World War II. With the war effectively cutting off Europe as a vacation destination, the Mexican Tourist Association worked with the US Office of Inter-American Affairs (OIAA) and commercial airlines to bring travelers to Mexico. First, however, they needed to counter popular perceptions of the country as a volatile and potentially dangerous hotbed of revolution and anti-Yankee sentiment. Companies relied on marketing that played on romanticized conceptions of exotic natives to promote ancient Indigenous ruins, while simultaneously presenting the country as modern and safe for vacationers. One advertisement for Aerovias Braniff that appeared in US travel magazines touted the "rich travel market in Mexico" with a picture of an Indigenous woman embroidering geometric patterns into a white cloth. CMA publicized the new airport it constructed in Mérida by touting its top-of-the-line amenities, its proximity to Mayan ruins, and its cutting-edge meteorological and radio communications systems. Once in Mérida, tourists could catch chartered flights directly to the pyramids of Chitzen Itzá. Such advertising strategies bore fruit, and tourism in the country increased by 7 percent over the course of the war. By 1946 Mexico had become a go-to destination for US tourists.[66]

Civil aviation experienced tremendous growth in the 1940s by several important measures. In terms of passengers, luggage, and distances flown, the industry made significant leaps immediately after Ávila Camacho took office. The war strengthened the relationship between government officials and business leaders and led to something of a technological renaissance. The CMA worked with the Ávila Camacho administration to develop a system of low-frequency radio navigation beacons along the country's east coast, and to advance meteorological science and technology. These breakthroughs, together with advances in training and greater access to equipment, allowed the industry to clear the material hurdles that had impeded it since the 1920s. The postwar tourism boom that began in 1943 appeared sustainable by 1945, and government officials and industrialists looked forward to cashing in on their success.[67] Clearing these hurdles allowed aviation to gain new, crucial momentum that carried the technology into peacetime.[68]

In the decades following the end of the war, policy makers viewed commercial aviation as essential to tourism and the country's overall development. During Ávila Camacho's presidency, aviation transformed from a symbolic technology, something that represented the country's aspirations as an industrializing nation but with little measurable effect on everyday life, into a foundational part of the postwar economy. The technology had built sufficient momentum so that experts and officials viewed the health of the industry as a bellwether of the economy. Nevertheless, while aviation remained central to policy makers' modernization programs, critics expressed anxieties about increased dependence on foreign tourists, inequality, and cultural exploitation.

MEASURING FLYING MACHINES AS A MEASURE OF MEXICO

Government leaders who supported the development of aeronautics, and the industry that it made possible, saw an opportunity to unite a country still in the grips of spasmodic upheaval in the decade following the outbreak of revolution. Aviation's unique potential to further the goals of national reconstruction by connecting the country physically, improving communications, aiding in the pacification of dissident forces, and speeding travel and trade, while avoiding the geographic obstacles that slowed construction of railways and highways, must have appeared miraculous to contemporary observers. Presidents like Álvaro Obregón, Plutarco Elías Calles, and Lázaro Cárdenas further believed that it offered a solution to a larger problem: how to preserve Mexico's unique identity and culture while allowing it to compete in a rapidly globalizing economy. Fulfilling aviation's promise nevertheless proved more difficult than officials initially had hoped.

While aviation was in the nascent stage of its development as a technology and industry, federal authorities succeeded in imprinting it with symbols and messages crafted by the country's intellectual and artistic elite. The technology thus reflected and contributed to a cultural revolution that reshaped national identity in the 1920s. This, combined with the sheer spectacle created by human flight, made aviation an effective medium for disseminating the values championed by the ruling party and stoking nationalist sentiment among the public. But this malleability was a double-edged sword. It required policy makers to try to protect the industry while the country attempted to restabilize politically and weather a global depression. These external conditions proved too difficult to overcome alone, leaving the door open to foreign interests, which came to dominate the industry much as they had dominated significant portions of the national economy during the latter half of the nineteenth century.

This outcome makes the patriotic image that officials and industrialists carefully cultivated on behalf of the national aviation program appear misleading. Rather than reconciling the conflicting impulses of the Second Industrial Revolution and the Mexican Revolution, aviation dressed the former in the clothes of the latter. That is, the public perception that the national aviation program represented progress for all Mexicans helped justify a development strategy that often worked against the interests of the working class, especially in rural communities. By the end of World War II aviation was no longer so readily susceptible to outside influence. It had matured into a crucial part of the government's postwar economic plan, which relied on continued industrialization and the growth of tourism, and favored the growing urban middle class over poor farmers and rural communities.

CHAPTER 9

A SOCIAL HISTORY OF URBAN EXPERTISE

Between Techno-bureaucratic Rule and the Right to the City in Twentieth-Century Mexico

MATTHEW VITZ

It is impossible to imagine a modern city without the interventions of a wide array of technical experts. So many of the comforts and amenities affluent urban inhabitants take for granted—the convenience of running water and sewerage, electricity, transportation infrastructure, and parks and green space—are the result of expertise accumulated and applied over time. Engineers, architects, doctors, and planners have sought to craft cities and direct urbanization in accord with the reigning paradigms of their day: for example, positivism, sanitarianism, functionalism, environmentalism, and so on. As state capacity increased and notions of urban modernity became tightly bound to technical expertise and capital accumulation during the nineteenth century, their efforts to create hygienic, aesthetically pleasing, and networked built environments—in short, the modern urban ideal—escalated.

This global city building phenomenon impacted every major metropolis, and it was rarely a matter of the Global South reflexively copying the North, as the champions of modernization theory had long asserted.[1] Ideas about the city circulated in ways not derivative of Europe and were adopted and adapted to fit local geographies and political imperatives. During the last decades of the nineteenth century in Mexico City, located in a closed basin prone to flooding, the liberal-positivist and Spenserian-influenced ruling class embraced technical and environmental knowledge of disease etiology and prevention, hydrology, engineering, and housing to fashion a habitable city. Urban experts—engineers, architects, and doctors—assumed the role of technocrats, joining the privileged positions held by the other economically minded *científicos* in Porfirio Díaz's government (1876–1911). Working in concert with private construction industries, they were charged with creating the networked and hygienic city, which stood as a source of state authority, a claim to legitimate rule. These experts, however, were not the sole authors of the modern city that they purported to be. Residents across Porfirian Mexico participated in the making of hygienic, networked urban spaces. Urban technological history has a social component of large sections of the population negotiating and clamoring for their inclusion in the modern city.

Once revolutionary violence waned in the early 1920s, much of the foundation of Mexico's one-party rule (1929–2000), particularly in its rapidly growing cities, rested on the promotion and maintenance of what I call a techno-bureaucratic state—a symbiotic albeit sometimes tense alliance among experts, state authorities, and private capital, who together held the capacity to fulfill the promise of the modern urban ideal. While technical experts tended to uphold their neutrality, officials politicized the new technological and networked city as a gift benevolently granted to the urban masses. Effaced in the discursive formation propagated by state (and private sector) experts was the notion that residents might themselves be legitimate co-constructors of the city and that under-regulated private capital might constrict state capacity to deliver its promises of adequate housing and sanitary infrastructure. Yet the urban popular classes, despite their general rejection of Porfirian politics, encouraged the gelling of this alliance. Popular struggles for the hygienic city in the 1920s and 1930s reshaped the urban landscape but rarely contested the charge experts held to remaking cities in the public interest. A new revolutionary state meant the triumph of expertise putatively working to benefit those who had not yet enjoyed the fruits of urban life.

Techno-bureaucratic urbanism of postrevolutionary Mexico, beholden to capitalist urbanization bristling with socio-spatial inequalities from real estate speculation and industrial development, was not smooth or uncontested. The techno-bureaucratic state laid claim to urban spaces and services capitalists had neglected—or could not profit from without state aid—but which they required for production. The material and ideological effects of these interventions in urban space set the parameters for urban politics and undergirded the clientelistic negotiations so thoroughly studied by sociologists and historians. Although housing and networked infrastructures served as emblems of responsible rule and imparted a modicum of urban political stability at mid-century, fractures in the legitimacy of techno-bureaucratic rule appeared by the late 1960s, coinciding with the global bankruptcy of high-state planning and technocratic utopianism. Urban residents, particularly on the poor periphery, built their own homes and infrastructures, and their city-making practices fostered an urban popular movement that transcended a mere desire to be included as benefactors of a benevolent state to demanding participation in decision making and design itself. The bottom-up challenges to Mexico's urban techno-bureaucracy culminated with the devastating 1985 earthquake in the nation's capital, which turned metaphorical fractures into physical ones—the terrifying testimony of an undemocratic alliance of bureaucrats with planners, engineers, and architects. The conflicts over the built environment helped end PRI rule.

Technocracy in Latin America tends to be associated with US-trained economists espousing neoliberal policies—free trade, privatization, deregulation, and monetarism—in the 1980s and 1990s and emboldened to undermine the power of career politicians connected to the old activist state. Yet historians have recently revealed a much longer history of expertise and its influence on economic policy, the built environment, and elite rule.[2] In fast-urbanizing Latin America—beginning with the export boom of the mid-nineteenth century—a host of engineers, architects, and health scientists worked tirelessly to refashion the built environment. My objective is to bring this important strand of scholarship on elite experts and the technological systems they created (with a focus on housing and sanitary services) into conversation with another that is generally treated separately: bottom-up political histories around what today would be dubbed "right to the city" movements.[3] In urbanizing Mexico, a triad of groups—urban experts, government bureaucrats, and nonelite residents (the small middle class and the much larger urban poor and working class)—wrangled over and sought to transform built environments. These urban actors sometimes conflicted with one another; other times they collaborated and even shared a conviction in technocratic rule. Understanding their evolving relationship in an unequal socio-spatial field sheds new light on the making and unmaking of the PRI state in particular and city-making practices more generally and may help us build potential futures for more just and sustainable cities.

TECHNOCRATS AND THE MODERN URBAN IDEAL

On the night of September 15, 1910, Mexico's centennial celebration, thousands of electric lights donning the buildings and streets of Mexico City's central business district shone brightly in an ostentatious display. In case anyone might have missed the message, the words "1910 progress" were spelled out with hundreds of lights on the Zócalo's main cathedral.[4] The completion of the hydroelectric dam at Necaxa, Puebla, in 1905 made electricity a powerful symbol of the grandeur of Porfirio Díaz's long-standing regime. The entirety of the centennial celebration in Mexico City boasted decades of Porfirian "order and progress," the positivist mantra wherein administration and the appropriate application of scientific knowledge would underwrite social improvement. The story of Mexican positivism is well known, so I won't belabor it here, but will draw the connections between its ethos of government and the rise of an urban expert elite.

Since independence Mexico's liberal leaders had dreamed of a more ardently capitalist Mexico governed by scientific knowledge instead of religion and tradition. They held a Whiggish view of a Mexico cloaked in superstition and crippled by Hispanic culture but with a rich endowment of natural resources primed for material progress once liberal trade, favorable European immigration policies, and a commitment to new technologies prevailed. Decades of devastating civil war had tempered their republican convictions and drew them to Comtean positivism. In Comte they found the elixir of progress. The technology and expert administration Mexico needed for its industrialization, urban growth, and increased health and worker productivity would be directed from above by Díaz's inner circle of científicos and a cadre of affiliated scientists and engineers. The liberal rule of representation was, by the late 1870s, supplanted by the rule of experts.

Technological networks required the specialized expertise of engineers working for capital, the state, or both. Foreign mining companies, manufacturers, agricultural interests, and railroad enterprises brought in their own engineers to implement new imported technologies. But liberal statesmen also understood the importance of cultivating domestic knowledge in mining, railway and hydraulic engineering, and health science to advance more public interests.[5] In 1867 Benito Juárez, under the direction of positivist philosopher Gabino Barreda, repurposed the old colonial mining college into the National Engineering School. Although instructors attempted to improve training—as discussed in this volume by Morelos and Escamilla—it remained modest throughout the revolution. Nonetheless, the school began graduating Mexico's first generation of engineers and architects, many of whom were instrumental in creating the nation's urban technological networks. They served on state-sanctioned technical commissions that oversaw and sometimes directly executed projects that came to embody the modern urban ideal in Mexico: electricity, water supply, sewerage and drainage, tramways, and planned subdivisions.

In Mexico City, Luis Espinosa, Roberto Gayol, and Manuel Marroquín y Rivera, graduates of the newly founded engineering school, devised the drainage of the basin, sewerage network, and water supply system, respectively. The three projects, completed between 1900 and 1912, cost an astounding 35 million pesos. The drainage was completed with the technology of Weetman Pearson's British company and overseen by the engineers of a state-run committee.[6] A French company installed the sewer network while being monitored by Gayol and his government committee. The state hired and funded Marroquín y Rivera to devise and build the water supply system, which relied on the Necaxa hydroelectric dam system for pumping. Although the electrical network was owned by Canadian and American interests, Mexican electrical engineers ensured the firm fulfilled its contract.[7]

These systems served as emblems of the hygienic city. They constituted Mexico's version of the technological sublime—the conviction that large technological systems would invariably improve human welfare. Scientist Jesús Galindo y Villa, public health official Eduardo Liceaga, and Federal District governor Pablo Escandón waxed panegyric about the new hydraulic infrastructure, which, they proclaimed, would forever banish flooding from the lake beds, clean city streets, and provide healthy drinking water to everyone.[8] Encrusted in their adulation of the networked city was a Promethean ideal regarding the capacity of science and technology to bend environments to fit capitalist urbanization. Good water was to be harnessed and brought into the modern home while bad water would be expelled.[9] It was in the making of the networked city that modern living became associated with control over nature: domesticated water, the conversion of darkness into light, and the hygienic home.[10]

Lest we fall into a Mexico City centrism, other major metropolises around the country also saw the engineering of new urban environments, often carried out by the same engineers who remade the capital. Marroquín y Rivera canalized the fetid San Juan de Dios River flowing through central Guadalajara, and city engineers updated the water system. In Veracruz the British magnate Weetman Pearson, who held a stake in the nearby oil industry, updated its water and drainage infrastructure and beautified the harbor boardwalk for the Porfirian elite. Mérida, in thrall of the henequen boom, also underwent improvements in its central districts, where wealthy plantation owners resided.[11] These initiatives did not approximate the investment or transformation in the nation's capital but neither were they merely derivative of a "Chilango blueprint."[12] Rather, they were contemporaneous and, in some instances, preceding developments in Mexico City.

Networked technologies and improved urban spaces did not reach everyone. Urban experts affirmed their role as public servants, but many held deep-seated prejudices against the lower classes, labeled insalubrious and uneducated. And the state's dual objectives of sanitizing cities and fostering real estate speculation and capitalist growth ran at loggerheads. Cash-strapped municipalities were often charged with the extension of sanitary infrastructure, circumscribing elites' ability to match their rhetoric of social improvement with investment. As a result, generally only wealthier residents, along with parts of a budding professional middle class, could boast of hygienic living spaces, paved streets, and electric lighting.

In Morelia the Porfirian technocratic promise rang hollow. The city council evoked the ideal of the hygienic, orderly city throughout the 1880s and 1890s, but financial constraints limited action. Residents, as historian Christina Jiménez has demonstrated, took the construction of the hygienic city into their own hands by submitting petitions with detailed blueprints and providing material and labor to complete sanitary infrastructure. Residents of all classes spoke the language of public health and displayed a modicum of technical knowledge, promising to keep sewage lines separate from water lines, use the correct pipe construction material "to hermetically seal [them] together," and ensure that sufficient drainage from homes would "allow the sewer to have a current of water which will constantly clean it out."[13] The technocratic ideal floundered in the face of popular organizing in this modern city that the Porfirian state espoused as its own conception and creation.

While some popular remonstrations in Morelia implicitly challenged the purportedly exclusive realm of technocrats to design the city, most citizen petitions across Mexico demanded top-down, technical solutions, sowing the seeds of dependence on the expert-led state. In the nation's capital, where greater state capacity enhanced technocratic urbanism, residents of poorly serviced *colonias* demanded the new networks be extended to their homes and streets.[14] The city's lower class, witnessing the transformations of their city, imbibed the proclamations of officials and subdivision developers regarding urban improvements and new infrastructures.[15] Technocratic urbanism embedded itself in popular culture at the turn of the century, a process by which the technical, rather than being evacuated of politics, was deeply politicized.[16]

It is thus no surprise that, in line with the Porfirian weltanschauung, inhabitants began to demand more technocracy to curtail disorder and insalubrity wherever they reared their ugly head. As fire service across urban Mexico became more technical and dependent on safety codes to respond to the growing risks of flammable structures, many people perceived long-standing community firefighting practices as obsolete. City residents demanded the new science of firefighting in their own neighborhoods.[17] When diseases like typhus or tuberculosis reached epidemic proportions, it was common for citizens to call for a "sanitary dictatorship" led by doctors and public health officials.[18] Technocratic urbanism was nonetheless very much a double-edged sword: it stoked public expectations of health and comfort on the one hand, but on the other, its reach too often stopped at the doors of affluence.

THE MEXICAN REVOLUTION AND THE MAKING OF THE TECHNO-BUREAUCRATIC STATE

The revolution of 1910 shook technocratic urbanism to its core, but the triumphant liberal-nationalist Venustiano Carranza buried the rural forces of Pancho Villa and Emiliano Zapata, who were wary of centralized authority. Carranza was no friend of Porfirian positivism, but he and his fellow Constitutionalists believed firmly in capitalist urbanization, which continued to require a heavy dose of technical expertise and elite guidance of the lower classes. Following the revolution, which reinvigorated democratic political persuasions, expertise contributed to the making of, and bound more strongly to, an urban public interest. The Constitution of 1917 mandated rights to health, hygiene, and home for Mexico's working classes. The Constitution and subsequent civil and health codes breathed new life into technocratic urbanism; technical experts and the state would be charged with guaranteeing the public interest against "reactionary" forces seeking to undermine revolutionary gains. And in metropolitan centers around the country, the rising expectations of revolutionary rights emboldened a restless populace to demand food safety, hygienic and affordable housing, and adequate public services.[19]

Nineteen twenty-two was a watershed year for the politics of housing and hygiene and therefore of urban expertise. In the spring and summer an untold number of tenants in cities across Mexico joined a rent strike initiated by anarcho-syndicalists in Veracruz. Following tenant unrest, Mexico City's water supply system broke down in the fall of that year, leaving hundreds of thousands of residents without the clean, piped water on which they depended. In its aftermath, urban residents, technical experts, and officials engaged at various levels in a struggle over who should create the city and govern its complex systems.

The rent strikers in Veracruz, Mexico City, and elsewhere placed a counterweight on the prevailing technocratic culture and laid bare an urbanization that had failed them. Not only were rents skyrocketing amid housing shortages, but existing units were derelict and bereft of sanitary connections. In a letter written from prison to his supporters, the anarchist leader of the Veracruz strikers, Herón Proal, elaborated a plan for urban governance wherein the union would requisition the tenements for public utility and "proceed with the installation of water and lighting." The union would also supply housing administrators to cover the costs of all "indispensable repairs such as roofs, floors, kitchens, bathrooms, [and] water closets."[20] Mexico City strikers, also numbering in the tens of thousands, set up committees of craftsmen to repair tenements, demonstrated in the streets and plazas, and confronted police executing eviction orders. For a brief moment in 1922, urban residents claimed the right to participatory administration and control over city spaces generally reserved for the state and capital.[21]

As police action and short-term concessions from landlords undermined the strike, authorities set aside tracts of land for working-class housing on the outskirts of Veracruz and Mexico City. Residents of cramped central districts occupied land that was intended to be state-sanctioned worker neighborhoods. In Mexico City Obregón hired architect Guillermo Zárraga to devise a modernist working-class settlement at the site of the old racetrack (Ex-Hipódromo de Peralvillo), replete with single-family homes and a full complement of services. Zárraga hoped this to be the first step in his larger vision of remaking the nation's capital, but funding even for this modest design dried up. Occupants proceeded to build their own homes in what was possibly Mexico's first *colonia popular*—a state-supported but resident-constructed subdivision.[22] Official acquiescence to self-help housing, which quickly became the most important form of popular housing, began in the wake of the 1922 rent strikes and the postrevolutionary elite's support for real estate capital and aversion to progressive planning ideas.

Even as poor urban residents began to rely on their own knowledge to build their homes, the technological sublime prevailed among a growing middle class weary from decades of civil strife that had wreaked havoc on many cities. The Stridentist movement, led by Manuel Maples Arce, invoked a techno-utopian future, while young writers like Salvador Novo effusively praised the modernizing urban landscape.[23] The prevailing confidence that a coterie of experts would collaborate with state and capital to improve welfare in Mexico's expanding cities remained palpable.

Even the crisis produced by the breakdown in water service galvanized support for appropriate technical expertise. In November 1922, when an electrical short disabled the pumps feeding water into the reservoir, water flooded the pump station and wrecked the system. The major dailies and leading hydraulic engineers denounced the penniless, and democratically elected, city council rather than question the possible design flaw by the system's builders, despite Marroquín y Rivera's unfortunate placement of the pumps in a floodable area of the station. Marroquín y Rivera himself accused the municipal government of negligence, while Modesto Rolland charged its members with graft, illegal sales of replacement parts, and obfuscation. Leading experts' ostensibly neutral reports lent credence to the press's anti-municipal crusade. City council members were in an unwinnable position, confronted with a crisis that was not entirely of their own making and battling a group of well-respected engineers policing the boundaries between what they believed was the superior technical realm and an inferior realm of electoral politics.[24]

Amid the worsening water scarcity and vituperative anti-municipal reporting, protests ensued. On November 30 thousands of workers marched to the seat of municipal governance where a riot broke out, the municipal registry was set ablaze, and the municipal guard opened fire on the crowd, killing several and wounding dozens. In the short term repression worked, and the city remained calm into December when full service resumed. But it was a public relations debacle for the municipal government. The press continued to assail the city council—the left-leaning *El Demócrata* called for its substitution with a corporatist-style "municipal soviet." Disgusted residents wrote Obregón to demand punitive measures and a technocratic city administration, from "a General Public Health Commission" to an "administrative advisory board."[25] Nonetheless, the city council escaped the crisis of 1922 buoyed by Obregón's party's victory in the municipal elections that followed the riot.

The status quo, however, was fragile. In 1928 Obregón issued a proposal to dissolve the municipal governments of the Federal District and replace them with a new Department of the Federal District (DDF), a centralized administration dependent on the executive branch. Politicians and a host of urban experts defended the proposal in language that invoked the discursive war on the city council during the water crisis. An administration in lockstep with Mexico's burgeoning planning culture and above the partisan messiness of municipal politics was essential to guide the city's growth and manage its complex technological systems. Rising Marxist labor leader Vicente Lombardo Toledano, fluent in urban planning debates, was a rare voice of dissent in a wilderness of technocratic centralism.[26] His alternative proposal of a democratic municipality encompassing the entire Basin of Mexico fell on deaf ears, and in January 1929 the DDF assumed governing responsibilities. To be sure, other factors motivated the disbanding of the Federal District's municipalities, but many powerful people held the firm belief that only administrative unity and the increased clout of experts would fulfill the promises of modern urbanism.

Rhetorical justifications notwithstanding, the ties the new cadre of planners, led by Columbia-trained Carlos Contreras, established with the DDF were tenuous. New planning and public works departments, for example, lacked funding. More often than not, city authorities used the language of technocracy as public relations stunts, such as when declaring a "water czar" to tackle continued water deficiencies.[27] In doing so they drank from the cup of popular political culture that looked to experts to purvey the fruits of urban modernity. Democratic municipal governance continued elsewhere—in name anyway—but events in Mexico City laid the ideological foundation for the techno-bureaucratic state in which professional bureaucrats established a working alliance with experts in public health, architecture, planning, and engineering, who served as intermediaries between the baleful effects of capitalist urbanization and a population seeking the hygienic and well-serviced city. In this vein, in the 1930s and early 1940s central government authorities set up expert advisory boards to govern urban infrastructure and centralized water service in San Luis Potosí, León, and Monterrey, among others.[28] Urban officials hired planners and engineers to slowly integrate new colonias populares into the networked city, often in response to a flare-up of urban protests demanding public services and land regularization, as was the case in late 1930s Mexico. These ephemeral urban social movements were Janus-faced: they challenged capitalist urbanization but also yielded to a state that used ad-hoc service extensions and property regularization to placate discontent. The messy and conflictive urban political culture of the 1920s and 1930s spawned techno-bureaucratic rule.

THE HALCYON YEARS OF TECHNO-BUREAUCRACY, 1940–1976

In Mexican history 1940 has become a hackneyed date. It signifies the dawn of a new era: the end of Cardenista reforms, the beginning of PRI corporatist hegemony (Pax Priísta), and the advent of the nation's so-called economic miracle. I employ the date here, with some trepidation, to signal an era of techno-bureaucracy in Mexico's bourgeoning cities. The techno-bureaucratic state stemmed from the immediate postrevolutionary period, and had even deeper roots in the Díaz regime. Yet after 1940 PRI authorities built up old and created new institutions that addressed specifically urban needs, and they sought out professional planners and engineers to legitimize their reign through the creation of the modern city. Agencies, commissions, and ministries associated with this task—Ministry of Hydraulic Resources, Ministry of Public Health and Welfare, Ministry of Communications and Public Works, National Housing and Public Works Bank (BNHUOP), Hydrological Commission of the Basin of Mexico, Department of the Federal District, and diverse planning commissions—were not always run by technocrats, but they hinged on the appropriate application of urban expertise above the political fray and, as was commonly invoked, in the name of the revolution.

Neither did experts always coalesce around PRI authority. Tensions abounded whenever career politicians felt their influence imperiled by the plans of technicians; the stagnated career of architect-planner Carlos Lazo after the completion of the monumental Ciudad Universitaria is but one example.[29] Yet urban experts were not constrained to employment by the state. They filled leading roles in the private sector as well: prominent architect of state-funded modernist housing Mario Pani had his own company, other architects and engineers worked in the private sector, and several joined to found their own, the high-profile Civil Engineers Association (ICA). Created in 1947, ICA executed dozens of urban construction projects across Mexico, and in the public imagination positioned itself as an anchor of modernization and social betterment alongside and in concert with the state.[30]

Nineteen forty marked a highly contested presidential election. The ideologically eclectic Juan Andreu Almazán absorbed much of the middle-class critique of Lázaro Cárdenas's revolutionary populism and likely carried Mexico City, Monterrey, and other urban centers. Shopkeepers, government employees, and mid-level urban professionals saw an unruly city in which the urban poor, as well as the rural peasantry, received special treatment and unwarranted political influence. They bemoaned the decrease in funding for urban services and viewed Almazán as their savior.[31]

Almazán's bid for the presidency failed, but authorities under the official party fine-tuned their flexible style of rule. Manuel Ávila Camacho prioritized questions of urban consumption, services, and welfare, in part to appease the Almazanista movement and to promote industrialization. Manufacturing promotion and the promotion of agribusiness over communal ejido agriculture turned Mexico City into a magnet for migrants. Incoming urban settlers established social networks to eke out a living, build their communities, and engage their political worlds.[32] And their political world revolved around the techno-bureaucratic state at city and federal levels and their perceived rights to the hygienic and affordable city. PRI authorities, committed to upgrading the urban built environment for political and economic reasons, set up water boards modeled on the Porfirian technical juntas and injected substantial funds to improve water, electricity, and sewer systems in municipalities around the country.[33]

Mexico City, whose geography combined with rapid growth challenged technical capacity, was the litmus test for the maintenance of the networked and hygienic city. The difficulties of meeting the needs of a growing population—the metro area grew from about 2 million in 1940 to about 9 million in 1970—was exacerbated by the city's ecological precariousness, located in a closed basin prone to flooding, water shortages, dust storms laced with pathogens, and land subsidence due to over-drafting of the underground aquifer. PRI rule rested on the foundation of applied urban expertise, in which hydraulic engineering and modernist housing served to assuage a disquieted urban populace.

Water service received the first injection of funding. In 1942, faced with another water crisis, the DDF and the federal government invested 226 million pesos to construct the first urban waterworks that transferred water from one watershed to another. Completed in 1951, it comprised an energy-intensive reservoir, canal, and pump system to bring the liquid to the city. According to water authorities, the system was a Mexican engineering marvel that "remedied 35 years of shortages."[34]

As city officials commemorated the Lerma engineering marvel, the drainage system repeatedly failed, flooding the historic center and surrounding neighborhoods. The destructive floods of 1951 and 1952 aggravated social tensions in a booming city with chronically deficient services, helping opposition candidate Miguel Henríquez Guzmán carry the Federal District in the contested 1952 presidential election.[35] In June of the previous year Miguel Alemán had created the Hydrological Commission of the Basin of Mexico and placed Ministry of Hydraulic Resources (SRH) head Adolfo Orive Alba at its helm. The engineer-dominated commission prescribed remedies to the city's multipronged water-related crisis—flooding, land subsidence, and the perennial dust bowl emanating from dried lake beds—that countered the conventional draft-and-drain paradigm and the PRI's industrial growth imperative with the purpose of restoring a lost hydrological balance.[36] The commission's idealism in the early 1950s gave way to realism several years later; new flood-prone development, new industrial sites tapping the aquifer, and the decaying condition of the fast-sinking Gran Canal required a massive technical fix that both reinforced conventional engineering and, like the Lerma project, served a legitimizing political function for disaffected city-center residents.[37] National Autonomous University of Mexico (UNAM) engineers teamed with SRH and DDF's Hydraulic Works Agency to design the monumental Deep Drainage System (El Sistema de Drenaje Profundo de la Ciudad de México), a network of tunnels and collectors that could withstand the basin's mushy soils susceptible to seismic activity and evacuate wastewater more quickly. Deep Drainage eclipsed the Lerma project in technical challenges and labor needs, and the DDF showcased the two works in similar fashion: as the solution to the flooding menace.[38]

Mexico's investments in modernist public housing, aimed to please the growing and increasingly influential urban middle class, complemented the grand hydraulic projects in strengthening the discursive and material ties between the PRI and technocratic urbanism. In coordination with BNHUOP, Mexican Institute of Social Security (IMSS), the Mexican institute for Social Security for State Workers (ISSSTE), and ICA, Mario Pani, Mexico's Le Corbusier, designed Mexico's first public housing complexes: Multifamilares Alemán, Multifamilaries Juárez, and Unidad Santa Fé in Mexico City, which housed thousands of state workers. State housing funding continued, averaging 20,800 residences per year between 1963 and 1970.[39] The overall effect was modest at best, but additional mammoth modernist housing reinforced the perception of state responsiveness to the aspiration of modern living. The headline of the late 1960s was the

Fig. 9.1. Rodrigo Moya, "Hipotecados," Tlatelolco, Ciudad de México, 1965.
Source: Archivo Fotográfico Rodrigo Moya.

Pani-designed Nonoalco-Tlatlelolco complex, the second largest housing complex in the Americas, a city within a city, with over one hundred multiuse buildings, large open spaces, recreational amenities, shopping facilities, and schools (see fig. 9.1). To make room for the new complex, officials eviscerated existing neighborhoods, which top officials smeared as morally depraved, prone to crime, and full of "hovels" without adequate hygiene.[40]

Architects and engineers, much like a large swath of the middle class drawn to antigrowth Federal District Regent Ernesto P. Uruchurtu, showed disdain for the self-built communities of Mexico's metropolises. These developments not only offended their aesthetic sensibilities but also threatened their power over city making and their perch as the privileged appliers of expertise.[41] Engineers on the Hydrological Commission, for instance, feared that poor migrants would occupy the federal zone of Lake Texcoco, disturbing the efficacy of the drainage works. The conservative architect Mauricio Gómez Mayorga groused that most residents lacked a "culture of habitation" (including those who had built their own homes) and derided the "vast arch of shacks growing at an alarming pace" around Mexico City in which "all hygiene, education, morality, and religion disintegrate." Gómez Mayorga preferred eradication and wanted to focus planning energies on beautifying the old central part of the city.[42] Most other architects of mid-century were neither as conservative nor as brazenly anti-poor in their thinking, but an implicit and coded language of contempt under-girded their modernist designs. Adolfo Zamora, chief of BNHUOP in the late 1940s, spoke of colonias populares as "mushrooms" that were quickly "strangling the city." The people there required a benevolent government teamed with experts to devise appropriate solutions.[43]

Pani's Santa Fé and Nonoalco-Tlatelolco complexes were among techno-bureaucracy's most visible "solutions" to the housing deficiency, a marker of sovereignty over urban space, and were proudly advertised as such in various media.[44] IMSS, which funded Santa Fé, promised not only hygienic modernist homes but health education, infant care, and recreation space. Residents of housing complexes often refashioned their own living and recreational spaces, but many cherished their living quarters, and broad sectors of the urban populace were enthralled by technocratic urbanism's promise to govern cities and forge a healthy environment.[45]

SELF-HELP HOUSING AND THE FISSURES IN TECHNO-BUREAUCRATIC RULE

Self-help housing revealed the limitations of authoritarian techno-bureaucracy as urban politicians sought to bring peripheral settlers into the state's fold. By the end of the 1960s Mexico faced a housing deficit of near-ly 6 million units. Despite the party's grandiloquence about a top-down techno-modernity, vast expanses of Mexico's cities were cocreated by resi-dents and public administration in a formalized set of political procedures first practiced after the rent strikes. Tenants from the overcrowded city cen-ters and recently arrived migrants settled thousands of colonias populares. Some settled on lands that urban developers had acquired, others gained access to lands from politicians who wanted to build their patronage, while many more organized to occupy ejido lands. Regardless, the settlers them-selves initiated construction, first to condition the land for sturdy construc-tion—a necessity on what tended to be precarious soils and then build their homes.

Urban theorist Jan Bazant Sánchez identified four types of self-built homes, which depended on a family's resources and the physical space where they were constructed. In all types families tended to build in stages, improving their homes when additional funds and labor became available. The more space and disposable income available, the nicer the home, as families could purchase sturdier materials, did not have to do backbreak-ing labor, or attempt technically sophisticated work on their own.[46] Home construction was generally a task completed by individual families, but communal reciprocity often obviated the need for hired labor in the most difficult tasks.[47] The acquisition of services soon followed. If the colonia was adjacent to an existing area with services, occupants would often illegally tap electricity lines. Many settlers, drawing on knowledge from managing water systems in their rural pueblos, might also coordinate to build an artesian well, clandestinely tap the municipal service, and perhaps even construct a makeshift network.[48]

Simultaneous to these joint technical efforts, they organized politically to gain legal land titles and formalized services. PRI authorities responded with selective interventions, new public policies, and a healthy dose of political control. New colonias had to demonstrate obedience while authorities, who posed as their saviors from predatory capitalists, employed planners and engineers to gradually integrate communities into the serviced city. Authorities also again drank from the cup of popular political culture dating back to the Porfiriato: *cooperación* (contribution). Formalized into official policy in 1931 through the Contribution Law, for which the Federal District government had its own officers to administer and direct works, the city compelled residents to fund and supply the labor to build their communities while its government provided the expertise.[49] Authorities also ensured that community political action remained within official bounds, localistic, and dispersed. These deft governing practices demobilized and placated the urban poor without requiring substantial resources. Nonetheless, if the residents of colonias populares on the peripheries of major cities wanted access to the networked city they saw advertised and promoted all around them, they could not rely on capital or the state alone, but rather on their own savings, their own physical strength, and sometimes their own technical expertise.

State authorities' rhetoric boasting of a technologically networked urban modernity within immediate reach belied the reality of rampant real estate speculation that structured housing scarcity and sanitary inequalities. The fluid and flimsy rule of techno-bureaucracy began bursting at the seams as the changing political and economic conditions of the 1970s combined with relentless population growth and the experience of self-help community construction to create an acute urban crisis.

THE CLEAVAGES OF TECHNO-BUREAUCRATIC RULE

Historians of Mexico are familiar with the narrative that marks the beginning of the crisis of PRI rule in 1968. Stultifying authoritarianism chafed against students and some workers nourished on the global radical politics of the 1960s. The demise of the economic miracle and the exhaustion of ISI in the early 1970s triggered inflation, and the 1985 earthquake exhibited the state's unresponsiveness to catastrophe in the middle of an economic crisis and awoke civil society. The politics of the built environment is mostly absent in this narrative. Urban spaces deteriorated during the 1970s and 1980s, revealing the inadequacy of the techno-bureaucratic state's quest to distribute access to healthy and comfortable living spaces. In 1970 over half the urban population lacked access to water and sewer systems, and adequate shelter remained scarce. Moreover, the rate of population growth of the nation's capital and several other major cities crept up, which further strained housing and services. Self-help housing and neighborhood building may have begun as a desperate practice against the exclusive nature of techno-bureaucratic governance, but it also encouraged residents to seek collective solutions, trust their own administrative practices, and devise their own expertise. City making, as James Holston argues for late twentieth-century Brazil, led working classes to develop a new idea of citizenship rights.[50]

A series of overlapping, if incongruous, urban social movements emerged in the 1970s and early 1980s among middle-class residents concerned about environmental pollution and among lower-class residents of forlorn colonias populares. Middle- and working-class actors who may have once acquiesced to the PRI's urban mission began questioning its direction. They not only demanded interventions from techno-bureaucracy but also proposed their own solutions by establishing grassroots organizations with the backing of dissident planners and engineers. Many of these organizations demanded a more radical notion of the right to the city: a seat in the planning rooms and at the decision-making table and substantive popular consultation.

If 1968 is the year most students of Mexico mention to denote the PRI's disintegrating legitimacy, 1969 was equally significant. In July of that year several dozen people met in a small self-built home in the rapidly growing Ciudad Nezahualcóyotl to discuss speculative land fraud. Independent of official institutions, they founded the Settlers Restoration Movement (Movimiento Restaurador de Colonos, MRC), proclaimed the land to be, in fact, property of the nation, and called for its expropriation.[51] Through militant activism that included large demonstrations, payment strikes, and defense against evictions, they won state subsidies to pay the developers for their land and the initiation of property regularizations and service extensions.[52] By 1974 the movement fractured between those who surrendered to state concessions and those who sought more radical measures, and it disintegrated. Although direct action was part of their repertoire, this was not a movement for participatory governance per se. It was, however, the first salvo in a battle over urban citizenship and the right to participate in the making of city space.

In the context of rural–urban migration, the new populist rhetoric of the Luis Echeverría administration (1970–1976), and the militant social Maoist activism of student leaders following the "mass line," cities such as Durango, Mexico City, and Monterrey saw an uptick in illicit land occupations and community building from scratch.[53] Some of these communities sought to negotiate within official channels, while others inaugurated a new urban leftism that forged solidarities with other working-class communities. During 1972 and 1973 residents in Durango formed the Independent Popular Union (Unión Popular Independiente, UPI); settlers in Chihuahua established the Popular Defense Committee (Comité de Defensa Popular); and residents of the northern and eastern "cinturón de la miseria" in Mexico City created the Union of Popular Settlements of the Valley of Mexico (Unión de Colonias Populares del Valle de México, UCP).

The purpose here is not to exhaustively list each urban popular organization but rather to give a sense of the movement's evolving relationship to the PRI's techno-bureaucracy. These organizations certainly had no qualms about demanding access to the networked city through the intervention of administrators and technical experts, but many were influenced by the Maoist objective to build popular power. And the more sophisticated urban popular administrative capacity became, the more coordinated their efforts were, the more independence they maintained, and the more they proffered a right to the city—not just of access but of participation in the planning and creating of urban space.

Shifts in planning policy likely also catalyzed a more direct confrontation with techno-bureaucracy. In global planning circles during the 1960s and 1970s, local and participatory action gained traction over high-modernist visions, at least ideologically if not in practice, and this tendency fit hand in glove with Echeverría's nominally democratic political opening to arrest mounting discontent.[54] In 1976 authorities launched the Law of Human Settlements, which stipulated the "consulta popular," and gave urban groups a say in discussions regarding conflict resolutions. Despite the language of inclusion, the law itself was drafted in closed-door meetings of technocrats and granted citizens merely the right to be heard, not necessarily the right to be listened to, much less the right to elaborate plans.[55] The 1980 Global Development Plan, put into effect by Echeverría's successor José López Portillo, also extoled popular participation but lacked the teeth to enforce it. Both the Planning Law and the Global Development Plan did normalize one way the urban populace engaged in city creation, however. Starting in 1975 federal, state, and municipal governments expanded the 1931 Contribution Law to assist settlers with the construction of their homes and networked infrastructures.[56]

Contemporaneous to these planning initiatives, popular movements pierced elite and technocratic planning spheres. In Monterrey chronic water shortages led to citywide rationing and ignited a mass protest movement in which tens of thousands participated. Some residents seized water-delivery trucks and kidnapped drivers, holding them until water officials directed piped water to their communities. Many others hit the streets and plazas to demonstrate against water shortages, echoes of 1922 Mexico City. These protests, like the MRC, were less about self-governance and participation in planning and more about demanding that the techno-bureaucratic state fulfill its promises. The protests worked, as the city strengthened its water supply system and implemented the plan "Agua para todos" to bring water to destitute areas. In the words of Vivienne Bennett, protesters "redefined citizens' rights . . . and while falling short of social transformation [they] challenged conventional planning in the water sector."[57]

The urban popular movement in Acapulco attacked techno-bureaucracy more directly. In 1980 Guerrero governor Ruben Figueroa schemed with Acapulco officials to resettle 125,000 residents of colonias populares from the touristy bay. The logic was environmental: reports by city and federal officials held colonias populares responsible for contaminating the bay, the city's natural amenity that attracted hundreds of thousands of tourists yearly. Residents formed the General Council of Colonias Populares of Acapulco to defend their neighborhoods, demand service extensions, and complete the land regularization process. They also insisted that the pollution derived from the hotels and the hundreds of boats that plied the bay, and they disputed the city's claim that residents living above 225 meters could not be administered municipal services, citing numerous examples of well-serviced homes and hotels above that line.[58] Pollution, according to government and nongovernmental reports, stemmed from all of the above sources, but the key point here is that Acapulqueños challenged official expertise and the exclusive city-making schemes that it sustained.[59] Despite demonstrations in the state capital supporting the residents, thousands were removed and resettled.[60]

The popular movements had the support of a bevy of antiestablishment planners working in academic institutions and NGOs. UNAM architecture students assisted settlers in *autoconstrucción* and in formulating plans for their makeshift colonias. Leaders of UCP worked with the Mexican Planning Society (Sociedad Mexicana de Planificación), and the umbrella organization, the National Coordinating Body of Mexico's Urban Popular Movements (CONAMUP), enjoyed the cooperation of some high-profile architects, including Pedro Moctezuma Barragán. New NGOs such as House and City (Casa y Ciudad) and Center of Housing and Urban Studies (Centro de la Vivienda y Estudios Urbanos, CENVI) assisted urban settlers with housing and neighborhood construction.[61] The UCP and CONAMUP, which comprised dozens of affiliate organizations across urban Mexico, worked tirelessly to construct a broad-based mass movement and organize diverse settlers of colonias populares in favor of a more participatory and egalitarian urbanization. The UCP built water networks, installed schools and medical clinics in several settlements, and set up consumer and housing cooperatives. They also demanded urbanization plans for all colonias populares be drafted in direct consultation with their organization and implored the state to carry out community-formulated urbanization projects—including electricity, transportation, and water lines.[62] CONAMUP's nationwide congresses led to the adoption of more general proposals. Beyond demanding regularization, rent reductions, and service connections, CONAMUP leadership devised their own projects, including "popular schools," health clinics, and consumer cooperatives.[63]

CONAMUP member organizations placed the environment on their agenda during the early 1980s, paralleling middle-class mobilizations over the capital's water and air pollution crisis. CONAMUP pleaded with the new Ministry of Urban Development and Ecology (SEDUE) to establish green areas and reforest hillsides around Mexico City's colonias populares, and members organized to stop the placement of a garbage dump near squatter settlers in Santiago Acahualtepec.[64] Organizers in the squatter settlements in the Ajusco mountains south of the city resisted a SEDUE expulsion decree and collaborated with radical planners and environmentalists to establish Productive Ecological Settlements (Colonias Ecológicas Productivas, CEPs) that protected woodlands, recycled rainwater for domestic use, and installed septic tanks to curb pollution.[65] These grassroots environmental planning initiatives undermined the state's rhetoric that removals were necessary to protect the fragile environment and exposed official environmentalism as an instrument of social control, not a neutral set of expert prescriptions.

Middle-class environmentalists and the urban popular movements converged on some other lightning-rod issues as well. Other examples include the successful campaign to close the polluting CROMATOS factory in the outskirts of Mexico City and mobilizations against PEMEX following the catastrophic explosion of the LP storage facility in San Juanico in 1984. However, middle-class environmental organizations were divided over the concerns of the urban poor, and moments of intersection were fleeting. Cultural stigmas on both sides—prevailing notions of the destructive poor on the one hand and Marxist dogma about class enemies on the other—no doubt played a role in dividing the two movements, but their separate struggles had more to do with their spatial and social positioning within the built environment. Middle-class environmental organizers saw a city on the verge of catastrophe, a once healthy space lost to overdevelopment, overpopulation and ill-advised hydraulic engineering. They envisaged a degraded urban ecology remedied only by their environmental expertise. Many considered peripheral settlements and industrial pollution as equally iniquitous—part of a homogenously conceived "overdevelopment" bereft of class antagonism. The urban popular movement, on the other hand, saw a city never made for them, a city that had to be re-created from the bottom up through self-help housing and independent organizing. A livable, hygienic city would require popular power, and while acknowledging the need for urban experts to realize this goal, the militant side of the urban popular movement tended to emphasize more systemic change and questioned the capacity of technical experts to advance democratic interests within the capitalist state.

"WE BECAME TECHNICAL EXPERTS"

By the summer of 1985 urban residents had been marching in the streets, refusing to pay mortgages and rents, forming independent organizations, and pressuring the techno-bureaucratic state to fulfill its promises. This was not merely a threat to the PRI's clientelism but very much also a bid to wrest the monopoly of city creation and administration from the state and its technocrats. Environmental activism and doomsday predictions were also on the rise in the mammoth capital—premonitions of a city on the brink, choked by pollution and people, running short of water, and susceptible to disasters. Moreover, the oil boom had gone bust, and a rising group of neoliberals had imposed a harsh austerity program. Then, on the morning of September 19, 1985, the earth released its fury. An 8.1 magnitude earthquake shook the loose, humid, and poorly consolidated soils of the Basin of Mexico for nearly two minutes. Plate tectonics and soil composition conspired with decades of construction to unleash widespread destruction in the central parts of Mexico City. An aftershock of 7.3 hit the following evening, adding to the catastrophe. Over ten thousand structures collapsed or were badly damaged and as many as forty-five thousand people perished.

The bankruptcy of technocratic urbanism that residents of peripheral colonias populares already felt suddenly pervaded the center of the nation's largest city. The earthquake, which, according to many analysts triggered the birth of Mexican civil society, was in reality its culmination following years of activism, grassroots administration, and working-class city building.[66] The act of nature catalyzed citizen participation in disaster relief, rescue, and rebuilding in central parts of the city that had been absent from the political ferment on the periphery. Once again, the crisis of the built environment defined relations between urban inhabitants and the techno-bureaucratic state.

The roar and screeches of collapsing steel, shattered glass, and crumbling concrete gave way to a deafening silence on the morning of September 19. The sounds of sirens were as rare as the cries for help were common. Much to the chagrin of quake victims, the government's response prioritized reviving production and protecting property. State employees immediately set out to reestablish telephone lines and bring back electricity, while emergency responders were sent to arrest looters and save machinery from collapsed factories.[67] Soldiers in the streets carried guns, not shovels, and President de la Madrid reportedly reassured international creditors that Mexico would not miss its debt payments.[68] Amid official negligence, thousands organized to remove rubble and aid victims.

On October 9 the PRI attempted to wrest control of the relief efforts from newly created citizen and community organizations. A panel of politicians, engineers, and other leaders inaugurated the Commission on Reconstruction in front of two thousand onlookers. Most of the speakers spilled platitudes about the perseverance of the national spirit without discussing concrete measures or revealing reconstruction plans.[69] Yet the earthquake strengthened thinking within and outside the state that Mexico City had grown too big and that political, economic, and ecological stability hinged on the deconcentration of industry and population. An outspoken member of the environmental organization Grupo de los Cien, spearheaded by poet Homero Aridjis, called for the ecological restoration of the city and the establishment of green spaces in the destroyed central areas. Similarly, the president of the Institute of Interdisciplinary Regional Planning and the vice president of ICA favored turning collapsed buildings into green areas.[70] For these thinkers, the quake was to be a catalyst of national regeneration through decentralization, ecological renewal, and the promotion of international tourism. These ideas percolated up to the highest levels; quake victims soon learned of plans to resettle them in the urban periphery to make room for a new *ciudad-jardín* or *museo-ciudad*.

The post-disaster dreamscapes of politicians and their planning allies faced stiff resistance. The collapse of the buildings and infrastructure that the city's techno-bureaucracy had constructed was as damaging to the PRI's image as its ineffective response during the search and rescue. The city of responsible expertise had become the city of corruption, tragedy, and ineptitude. The Nuevo Leon building in the Tlatelolco housing complex, the Juárez and Miguel Alemán housing complexes, five major hospitals, the Ministry of Labor, the Ministry of Communications, and the Ministry of Commerce all had crumbled. Other major government buildings and installations had also collapsed or were severely damaged—including forty-two other towers in the Tlatelolco complex. The parastatal telephone company's major telephone exchange building and several stations of the Federal Electricity Commission lay in ruins, leaving the capital incommunicado and dark.[71] Decades-old water mains and sewer collectors burst. Furthermore, middle-class condominiums and apartments in neighborhoods such as La Roma, La Condesa, and La Juárez fell or were irreparably damaged, and older, already dilapidated tenement buildings in central neighborhoods suffered extraordinary damage. In the words of Diane Davis, "For days and sometimes weeks or longer, hundreds of thousands of people had no homes, no work, no transportation, no food, no water, no telephones, no hospitals to visit for treatment of the wounded."[72]

When public attention shifted to reconstruction, anger against leading technocrats and bureaucrats mounted. Residents of Pani's Tlatelolco housing complex had for years implored government officials to address much-needed repairs and maintenance. The Nuevo Leon building showed structural deficiencies in the early 1980s, implicating Pani directly. By 1985 responsibility for upkeep lay with housing authorities, who when confronted by residents "displayed blueprints and sketches, using technical terms . . . telling us in sum that 'The Nuevo Leon is the safest building not in Tlatelolco, but in Mexico City.'"[73] Post-quake studies of the Tlaltelolco complex proved residents' earlier distrust of housing authorities correct.

Authorities tended to blame the residents themselves or "nature"—the earth shook so buildings fell—but most victims knew better.[74] They identified a web of complicity involving private construction firms, tenement landlords, state officials, and engineers and architects. Miguel de la Madrid charged SEDUE's head, architect Guillermo Carrillo Arena, with the reconstruction campaign, a move that drew the ire of the growing post-quake citizens' movement. Activists were perturbed that one of the primary individuals "responsible for our tragedy" would lead reconstruction.[75] Elena Poniatowska, author of the classic journalistic account of the earthquake, captured the tenor of the city's inhabitants: "The hands of assassins that got the profits should not be the hands that give back to us what we have lost."[76] To add insult to injury, Carrillo Arena's demeanor in meetings seemed disdainful, patronizing, and imperious.[77] At a march of thirty thousand to the president's home, chants railed against him, "Carrillo Arena, watch your tail, the people want you to go to jail."[78]

De la Madrid soon thereafter removed Carrillo Arena and placed a leading academic sympathetic with the citizens' movement in a key liaison position. This signaled a shift in how the government approached the citizens' movement.[79] Although not a single politician, architect, or engineer was charged for corruption or criminal negligence, the PRI was compelled to abandon its plans for resettlement.

The movement transcended scathing language. Victims and dissident experts devised their own proposals for reconstruction that militated against corruption. In October resident organizations formed the Overall Coordinating Committee of Disaster Victims (Coordinadora Única de Damnificados, CUD), which served as a clearinghouse for demands and public outreach. Although CUD and CONAMUP leadership discussed the possibility of a broad-based urban coalition, political-ideological disagreements hampered their working relationship.[80] The earthquake quickened the devolution of authority, and to some degree expertise itself, to the populace organizing to (re)construct the city.

The CUD demanded active involvement in reconstruction. Many members shunned government assistance and called on uncompromised foreign experts to determine the habitability of damaged buildings.[81] They requested the expropriation of destroyed housing and the right to oversee their reconstruction in situ. Victims, who had already made a tradition out of making repairs to derelict tenements before the quake, enlisted urban experts to re-create homes and neighborhoods out of the rubble.[82]

Authorities ceded ground as thousands took to the streets—including previously complacent middle-class residents of Tlatelolco. The president established an independent technical committee composed of engineering and architecture faculty to carry out inspections, expropriated thousands of damaged properties, and placed popular housing reconstruction at the center of the government's program. In May 1986 the government, CUD leaders, and myriad other professional organizations signed the Pact of Democratic Agreement on Reconstruction. According to CUD representative Cuauhtémoc Abarca, the pact "marked a milestone in the sense that . . . the state accepts the plurality of society and recognizes social participation."[83] The pact mandated that the government coordinate with community organizations and their architect partners to design and construct "dignified housing" resistant to earthquakes and with adequate sanitary services.[84]

The participatory planning outlined in the pact had teeth, and it was, in many ways, an ex-post-facto formalization of popular practice that evolved after September 19. Residents had sometimes already assumed the role of rebuilder. Even before the president initiated the expropriations, for example, residents worked with Casa y Ciudad to demolish buildings and devise prototypes. Casa y Ciudad, CENVI, and several CUD-affiliated organizations built dozens of homes, often with government financial aid.[85]

This independent reconstruction, akin to what had been occurring in peripheral settlements, was not universal; most organizations cooperated with the state. Nevertheless, when the federal state's Housing Renovation Program and SEDUE entered destroyed neighborhoods to offer technical advice, machinery, and financial aid, citizen organizations had already cleared rubble, devised housing blueprints, and fought, often successfully, to bring them to fruition.[86] Regardless of the degree to which cooperation with the state was pursued, "the fundamental question," as activist Miguel Armas affirmed, "was the construction of the city under parameters different from capital" and its backers in the PRI.[87]

Yet the more the citizens' movement addressed structural inequalities rooted in neoliberal capitalist urbanization, the more Sisyphean the struggle became. The movement for the right to the city gradually transformed into the movement for the neighborhood because of PRI co-optation and the gradual democratic opening in Mexico City and beyond.[88] Even local achievements were adroitly redirected by the neoliberal state in the long run. The rebuilding of working-class residences in situ came at the cost of rent control, which opened the city center to the speculative land market and urban gentrification. The renewal of the historic center, initiated by left-center mayor Andrés Manuel López Obrador and business magnate Carlos Slim in the early 2000s, originated in the post-quake rebuild. And while the confluent urban popular movements, centered in Mexico City after 1985, helped achieve electoral democracy, first in the capital (1997) and then nationwide (2000), their bid to restructure urban space and participate in planning foundered. The owl of Minerva—the wisdom of participatory planning gained through grassroots mobilization—flew at dusk.

Neoliberal orthodoxy supplanted the old alliance of modernist planning and an activist state promising urban modernity. Big engineering and architectural projects have continued, but despite the usual rhetoric of community improvement, they increasingly respond above all to the logic of capital accumulation and the imperative to brand cities as aesthetically pleasing spaces of consumption.[89] On the centennial of the Mexican Revolution, a revived urban popular movement pushed the democratic Mexico City government to enact a "Right to the City" charter, which spawned popular housing and community development initiatives. Such measures are redolent of policies from the late 1970s and 1980s wherein the state appropriated *auto-construcción* as its own while championing popular empowerment. Whereas undoubtedly these recent victories represent another bubbling-up of democratic city making, the more substantive right to the city remains distant. A new technocratic urbanism has germinated, one that speaks the language of localism and democracy but has confined the most salient decisions regarding city administration within a neoliberal framework.

NOTES

INTRODUCTION: ENGINEERING AND TECHNOCRATIC VISIONS IN MEXICO

1. Many historians refer to the network of industrial developments between the mid-nineteenth and mid-twentieth century as the Second Industrial Revolution or the second phase of the Industrial Revolution, but it is problematic when applied to Mexico. I am using it as shorthand for the wave of industrial transformations that took place globally and within Mexico during this time. Although machines and practices from the early stages of the Industrial Revolution (1750–1850), such as the use of steam engines in mining and textiles, existed in Mexico, its economy was still powered overwhelmingly by solar, water, and muscle power. In that sense, the Second Industrial Revolution, which included the development of the Bessemer process for steel production, railroads, telegraphs, telephones, interchangeable parts, modern sewage systems, automobiles, and airplanes, was Mexico's first period of significant industrialization. See David Pretel and Lino Camprubí, *Technology and Globalisation: Networks of Experts in World History* (London: Palgrave Macmillan, 2018); Peter N. Stearns, *The Industrial Revolution in World History*, 4th ed. (New York: Routledge, 2012); Germán Vergara, *Fueling Mexico: Energy and Environment, 1850–1950* (New York: Cambridge University Press, 2021); Aurora Gómez-Galvarriato, *Industry & Revolution: Social and Economic Change in the Orizaba Valley, Mexico* (Cambridge, MA: Harvard University Press, 2017).

2. For those interested in the history of engineering, some solid works include Richard Shelton Kirby et al., *Engineering in History* (1956; New York: Dover Publications, 1990); Ken Alder, *Engineering the Revolution: Arms and Enlightenment in*

France, 1763–1815 (Chicago: University of Chicago Press, 1997); John Rae and Rudi Volti, *The Engineer in History*, rev. ed. (New York: Peter Lang, 2001); David Cannadine, "Engineering History, or the History of Engineering? Rewriting the Technological Past," *Transactions of the Newcomen Society* 74 (2004): 163–80; Antoine Picon, "Engineers and Engineering History: Problems and Perspectives," *History and Technology* 20, no. 4 (2004): 421–36; Ben Marsden and Crosbie Smith, *Engineering Empires: A Cultural History of Technology in Nineteenth-Century Britain* (London: Palgrave Macmillan, 2005); David Muir-Wood, *Civil Engineering: A Very Short Introduction* (New York: Oxford University Press, 2012).

3. For more on experts, technology, and mediation, see Bruno Latour, *Reassembling the Social: An Introduction to Actor-Network-Theory* (New York: Oxford University Press, 2005). Exploring engineers as mediators and "middle people" also fits into Thomas J. Misa's discussion on using middle-level or "meso" studies as a way to better understand the discrepancies between technological determinism-leaning macro studies and societally contingent micro studies of technology. See Thomas J. Misa, "Retrieving Sociotechnical Change from Technological Determinism," in *Does Technology Drive History? The Dilemma of Technological Determinism*, ed. by Merritt Roe Smith and Leo Marx (Cambridge, MA: MIT Press, 2001), 115–42; Paul N. Edwards argues that infrastructure can also link macro and micro scales of study in "Infrastructure and Modernity: Force, Time, and Social Organization in the History of Sociotechnical Systems," in *Modernity and Technology*, ed. Thomas J. Misa, Philip Brey, and Andrew Feenberg (Cambridge, MA: MIT Press, 2003), 185–226.

4. Rae and Volti, *Engineer in History*, 2–4.

5. Chinampa was a form of farming in central Mexican lakes that used rectangular, often insular plots made from lake-bottom mud and intersected by canals of lake water. For more on hydraulic engineering and the building of docks and roads in Aztec Mexico, see Ross Hassig, *Trade, Tribute, and Transportation: The Sixteenth-Century Political Economy of the Valley of Mexico* (Norman: University of Oklahoma Press, 1985), chaps. 2 and 3. For works on land and water manipulation during the colonial period, see James Lockhart, *The Nahuas after the Conquest: A Social and Cultural History of the Indians of Central Mexico, Sixteenth through Eighteenth Century* (Stanford, CA: Stanford University Press, 1992), chap. 5; Vera S. Candiani, *Dreaming of Dry Land: Environmental Transformation in Colonial Mexico City* (Stanford, CA: Stanford University Press, 2014); Rani Alexander, ed., *Technology and Tradition in Mesoamerica after the Spanish Invasion: Archaeological Perspectives* (Santa Fe: University of New Mexico Press, 2019).

6. Juan D. Lindau, "Technocrats and Mexico's Political Elite," *Political Science Quarterly* 111, no. 2 (Summer 1996): 295–322; Sergio Aguayo, *Myths and [Mis] Perceptions: Changing U.S. Elite Visions of Mexico* (Berkeley: University of California Press, 1998), 201–11; Miguel Ángel Centeno and Patricio Silva, *The Politics of Expertise in Latin America* (Houndmills, UK: Palgrave, 1998); Miguel

Ángel Centeno, *Democracy within Reason: Technocratic Revolution in Mexico*, 2nd ed. (University Park: Pennsylvania State Press, 1997).

7. Roderic A. Camp, *Mexico's Mandarins: Crafting a Power Elite for the Twenty-First Century* (Berkeley: University of California Press, 2002). Also see Roderic A. Camp, "The Political Technocrat in the Mexico and the Survival of the Political System," *Latin American Research Review* 20, no. 1 (1985): 97–118. Camp tends to focus on economists, but he also includes other experts within the term "technocrat," including engineers, architects, and medical doctors.

8. Sarah Babb, *Managing Mexico: Neoliberals and the Globalization of Economic Expertise* (Princeton, NJ: Princeton University Press, 2011), 171.

9. Lewis Mumford, *Technics and Civilization* (1934; Chicago: University of Chicago Press, 2010); Mumford, *The Conduct of Life* (New York: Houghton Mifflin, 1951); Mumford, *Technics and Human Development: The Myth of the Machine*, vol. 1 (Boston: Mariner Books, 1971); Mumford, *The Pentagon of Power: The Myth and the Machine*, vol. 2 (Boston: Mariner Books, 1975); Herbert Marcuse, *One-Dimensional Man: Studies in the Ideology of Advanced Industrial Society* (Boston: Beacon Press Books, 1964); Jacques Ellul, *The Technological Society* (New York: Vintage Books, 1967); Theodore Roszak, *The Making of a Counter Culture: Reflections of the Technocratic Society and Its Youthful Opposition* (Berkeley: University of California Press, 1968). The notion of rationalization is based on the idea that the growth of science, technology, and bureaucracy focused on scientific principles and technical expertise would increasingly secularize, depersonalize, and drive networks of economic and political power. It was popularized by sociologist Max Weber. It is problematic in its positivistic sense of progression and it fails to consider resistance to this trend or the persistent avarice of humanity. However, rationalization holds an explanatory appeal in a world in which laws, scientific standards, and even cultural norms become marked by increased uniformity. See Max Weber, *Economy and Society*, 2 vols. (Oakland: University of California Press, 2013). This work was originally published posthumously in the early 1920s. On standardization and uniformity, see C. A. Bailey, *The Birth of the Modern World, 1780–1914* (Malden, MA: Blackwell, 2004).

10. Magali Sarfatti Larson, "Notes on Technocracy: Some Problems of Theory, Ideology, and Power," *Berkeley Journal of Sociology* 17 (1972–1973): 4.

11. Roszak, *Making of a Counter Culture*, 7–8.

12. James C. Scott, *Seeing Like a State: How Certain Schemes to Improve the Human Condition Have Failed* (New Haven, CT: Yale University Press, 1998), 3–4.

13. Daniele Caramani, "Introduction: The Technocratic Challenge to Democracy," *The Technocratic Challenge to Democracy*, ed. Eri Bertsou and Daniele Caramani (New York: Routledge, 2020), 3–4.

14. Caramani, "Introduction," 4.

15. The resulting single-party state that came to dominate Mexico from the 1930s to 2000 is more representative of a soft authoritarianism than an actual

representative democracy, but many of the technocrats who joined the revolution fought for a representative democracy.

16. This fits historical studies about technocrats in other parts of Latin America. See Eve Buckley, *Technocrats and the Politics of Drought and Development in Twentieth-Century Brazil* (Chapel Hill: University of North Carolina Press, 2017).

17. It is not surprising that the periodization for this edited volume dovetails with that of the first book to place the transition to fossil fuel energy at the center of Mexican history. Engineers and technocrats were integral parts of that story. See Vergara, *Fueling Mexico*. For more on the history of energy in Latin America, also see Amelia Kiddle, ed., *Energy in the Americas: Critical Reflections on Energy and History* (Calgary: Calgary University Press, 2021).

18. For more on experts in comparative and global history, see Pretel and Camprubí, *Technology and Globalisation*. For studies of experts in Cold War Latin America, see Andra B. Chasteen and Timothy W. Lorek, eds., *Itineraries of Expertise: Science, Technology, and Environment in Latin America's Long Cold War* (Pittsburgh: University of Pittsburgh Press, 2020).

19. Hassig, *Trade, Tribute, and Transportation*.

20. Peter W. Rees, "Origins of Colonial Transportation in Mexico," *Geographical Review* 65, no. 3 (July 1975): 323–34.

21. See Juan C. Lucerna, "*De Criollos a Mexicanos*: Engineers' Identity and the Construction of Mexico," *History and Technology* 23, no. 3 (September 2007): 275–88.

22. "Casta" was a Spanish colonial term for people of mixed races.

23. Lucerna, "*De Criollos a Mexicanos*," 275–76.

24. The university I work for, Arkansas State University, can be seen as a part of this process with the establishment of their Querétaro campus, in partnership with local Mexican entrepreneurs. The campus focuses on business, engineering, and transnational connections.

25. Lucerna, "*De Criollos a Mexicanos*," 277–81; J. Justin Castro, *Apostle of Progress: Modesto C. Rolland, Global Progressivism, and the Engineering of Revolutionary Mexico* (Lincoln: University of Nebraska Press, 2019), 8–11.

26. This volume provides a good comparison for studying technocracy during this period with Timothy Mitchell, *Rule of Experts: Egypt, Techno-politics, Modernity* (Berkeley: University of California Press, 2002).

27. Castro, *Apostle of Progress*, 17–20.

28. Historians often give different dates for the end of the revolutionary period. The military contests for power mainly occurred from 1910–1920, though revolts continued well into the 1930s. The most revolutionary policies, for example, massive land redistribution programs and the Mexican oil expropriation, occurred during the presidency of Lázaro Cárdenas (1934–1940), though they carried on in more moderate forms thereafter. I selected 1946 as the ending date

because it marks the presidential election of Miguel Alemán (1946–1952), who was the first civilian president to come to power after the revolution. That year the revolutionary party took on its final and institutionalized name: the Partido Revolucionario Intitucional (PRI). By this time a number of Mexican technical institutes shifted engineering education, aligning more with private, transnational enterprise.

29. This influence was by no means one-directional, but US companies used US engineers in most of their mining and railroad operations in Mexico. Many US engineering texts were used in Mexican universities. And some Mexican engineers received their education at MIT and other US engineering schools. An author who looks at the influence of the US Department of Interior and a specialist on US expansionism is Megan Black, *The Global Interior: Mining Frontiers and American Power* (Cambridge, MA: Harvard University Press, 2018).

30. US Census Bureau, "Top Trading Partners—December 2019," https://www.census.gov/foreign-trade/statistics/highlights/top/top1912yr.html.

31. See Tanalís Padilla, *Unintended Lessons of Revolution: Student Teachers and Political Radicalism in Twentieth-Century Mexico* (Durham, NC: Duke University Press, 2021), n. 33.

32. Lefebvre, *Key Writings* (New York: Bloomsbury Academic, 2017); Peter Marcuse, "From Critical Urban Theory to the Right to the City," *City* 13 (2009): 185–97; David Harvey, "The Right to the City," *New Left Review* 53 (September-October 2008): 23–40; Clara Izábal, ed., *Ordinary Places, Extraordinary Events: Citizenship, Democracy and Public Space in Latin America* (New York: Routledge, 2008). Examples of environment and technology studies include Sara B. Pritchard and Carl A. Zimring, *Technology and the Environment* (Baltimore: Johns Hopkins University Press, 2020); Vergara, *Fueling Mexico*; Mikael D. Wolfe, *Watering the Revolution: An Environmental and Technological History of Agrarian Reform in Mexico* (Durham, NC: Duke University Press, 2017). For influential works about the politics of development and problems and failures in modern development schemes, see Scott, *Seeing like a State*; Mitchell, *Rule of Experts*; Buckley, *Technocrats and the Politics*.

33. Vols. 1–15 of *Quipu* can be accessed online at http://www.revistaquipu.com/Sub1/?page_id=1133&fbclid=IwAR32cZ83aKVBZkpsXNE4ERYyDIbvs0EwkJ0eKTFLVj8LbtvNmMCZRrLiDGY.

34. For example, José María López Piñero, *Ciencia y técnica el la sociedad Española de los siglos XVI y XVII* (Barcelona: Labor, 1979). For more recent takes on this topic, see Jorge Cañizares-Esguerra, *Nature, Empire, and Nation: Explorations of the History of Science in the Iberian World* (Stanford, CA: Stanford University Press, 2006); and Antonio Barrera-Osorio and Mauricio Nieto Olarte, "Ciancia, tecnología, sabers locales e imperio en el mundo Atlántico, siglos XV–XIX," *Historia Crítica* 3 (2019): 3–20. Many of the first issues of *Quipu* leaned heavily in favor of topics about science in colonial Latin America.

35. David Wade Champers, "Review: *Quipu: Revista Latinoamericana de Historia de la Ciencias y Tecnología* by Juan José Saldaña," *Isis* 82, no. 2 (June 1991): 323–24.

36. A now-classic work on professions is Andrew Abbott, *The System of Professions: An Essay on the Division of Expert Labor* (Chicago: University of Chicago Press, 1988).

37. Francisco Arce Gurza et al., *Historia de las profesiones en México* (Mexico City: Colegio de México, 1982).

38. José Ramon de Ibarrola, *Apuntes sobre el desarollo sobre el ingeniería en México y le educación en ingeniería* (Mexico City: Tipografía de la F. Diaz de Leon, 1911); Daniel Reséndiz Nuñez, *La investigación en ingeniería: Consideraciones sobre su historia en México* (Mexico City: Instituto de Ingeniería, UNAM, 1979): Maria de La Paz Ramos Lara and Juan José Saldaña, "Del Colegio de Minería de México a la Escuela Nacional de Ingenieros," *Quipu* 13, no. 1 (January–April 2000): 105–26; Libertad Díaz Molina and Juan José Saldaña "Los ingenieros mexicanos y la reglamentación de la industria eléctrica, 1923–1933," *Quipu* 16, no. 1 (January–April 2013): 101–24.

39. Mílada Bazant, "Estudiantes mexicanos en el extranjero: El caso de los hermanos Urquidi," *Historia Mexicana* 36, no. 4 (April–June 1987): 739–58; Mílanda Bazant, "La enseñanza y la práctica de la Ingenería durante el porfiriato," en *La educación en la historia de México*, ed. Josefina Zoraida Vázquez (Mexico City: Colegio de México, 1992), 167–210.

40. Ivan San Martín Córdova, ed., *Ingenieros de profesión, arquitectos de vocación: Venticinco protagonistas en la Arquitectura Mexicana del siglo xx* (Mexico City: Universidad Autónoma Nacional de México, 2020).

41. Raúl Domínguez Martínez, *La ingeniería civil en México, 1900–1940: Análisis histórico de los factores de su desarrollo* (Mexico City: Universidad Autónoma Nacional de México, 2013). One of the better works on the earliest years of civil engineering and architecture in Mexico, that is during the 1850s and 1860s, is Leopoldo Rodríguez Morales, *El campo del constructor en el siglo XIX: De la certificación institucional a la esfera pública en la ciudad de México* (Mexico City: Instituto Nacional de Antropología e Historia, 2012).

42. Martínez, *La ingeniería civil en México*, 14. Translation is my own.

43. Priscilla Connolly, *El contratista de Don Porfirio: Obras públicas, deuda y desarrollo desigual* (Mexico City: Fondo de Cultura Económica, 1996).

44. Luis Aboites Aguila, *La decadencia del agua de la nación: Estudio sobre desigualidad social y cambio politico en México, segundo mitad del siglo xx* (Mexico City: El Colegio de México, 2009). For examination of conflicts over water in urban Mexico in the late twentieth century and early twenty-first century, see María Luisa Torregrosa, ed., *El conflicto de del agua: Política, gestion, resistencia y demanda social* (Mexico City: FLASCO, 2017).

45. Theodore Veblen, *The Theory of the Leisure Class* (New York: Macmil-

lan, 1899; New York: Dover Publications, 1994); Veblen, *The Engineers and the Price System* (Kitchener, ON: Batoche Books, 2001) [originally published in 1921]; Mumford, *Technics and Civilization*.

46. Organizational notes authored by Melvin Kranzberg, "The Society for the History of Technology: A Brief History," *Technology & Culture* 1, no. 1 (Winter 1959): 106–8. A solid history of SHOT and the historiography of STS up until 2002 can be found in John M. Staudenmaier, "Rationality, Agency, Contingency: Recent Trends in the History of Technology," *Reviews in American History* 30, no. 1 (March 2002): 168–81.

47. For the former, see Wiebe E. Bijker, Thomas P. Hughes, and Trevor Pinch, eds., *The Social Construction of Technological Systems: New Directions in the Sociology and History of Technology*, anniversary ed. (Cambridge, MA: MIT Press, 2012); Merritt Roe Smith and Leo Marx, eds., *Does Technology Drive History? The Dilemma of Technological Determinism* (Cambridge, MA: MIT Press, 1994). For the latter, see Latour, *Reassembling the Social*. Examples of philosophical works on the interplay between society and technology include Misa, Brey, and Feenberg, *Modernity and Technology*; Sheila Jasanoff, *The Ethics of Invention: Technology and the Human Future* (New York: W. W. Norton, 2016); Andrew Feenberg, *Technosystem: The Social Life of Reason* (Cambridge, MA: Harvard University Press, 2017); Langdon Winner, *The Whale and the Reactor: A Search for Limits in the Age of High Technology*, 2nd ed. (Chicago: University of Chicago Press, 2020).

48. This quote comes from Edin Medina's MIT website, https://edenmedina. mit.edu.

49. María M. Portuondo, *The Spanish Disquiet: The Biblical Natural Philosophy of Benito Arias Montano* (Chicago: University of Chicago Press, 2019); María M. Portuondo, *Secret Science: Spanish Cosmography and the New World* (Chicago: University of Chicago Press, 2009); Lucerna, "*De Criollos a Mexicanos*," 275–88.

50. For more on this, see J. Justin Castro, "History of Technology and Society in Nineteenth and Twentieth Century Latin America," *History Compass* (January 2020): 1–11, https://onlinelibrary.wiley.com/doi/abs/10.1111/hic3.12609.

51. Examples include Diana J. Montaño, *Electrifying Mexico: Technology and the Transformation of Mexico City* (Austin: University of Texas Press, 2021); Rocio Gomez, *Silver Veins, Dusty Lungs: Mining, Water, and Public Health in Zacatecas, 1835–1946* (Lincoln: University of Nebraska Press, 2020); Michael K. Bess, *Routes of Compromise: Building Roads and Shaping the Nation in Mexico, 1917–1952* (Lincoln: University of Nebraska, 2017); Anna Rose Alexander, *City on Fire: Technology, Social Change, and the Hazards of Progress in Mexico City, 1860–1910* (Pittsburgh: University of Pittsburgh Press, 2016); J. Justin Castro, *Radio in Revolution: Wireless Technology and State Power in Mexico, 1897–1938* (Lincoln: University of Nebraska Press, 2016); Rob Alegre, *Railroad Radicals: Gender, Class, and Memory in Cold War Mexico* (Lincoln: University of Nebraska Press, 2014); Joanne Hershfield, "Domestic Technologies: Gender, Technology, and Mexican House-

wives, 1930–1950," in *Technology and Culture in Twentieth-Century Mexico*, ed. Araceli Tinajero and J. Brian Freeman (Tuscaloosa: University of Alabama, 2013); Michael Matthews, *The Civilizing Machine: A Cultural History of Mexican Railroads, 1876–1910* (Lincoln: University of Nebraska Press, 2013); Matthew Vitz, *A City on a Lake: Urban Political Ecology and the Growth of Mexico City* (Durham, NC: Duke University Press, 2018). One of the more important books on technology and race in twentieth-century Mexico is David Dalton, *Mestizo Modernity: Race, Technology, and the Body in Postrevolutionary Mexico* (Gainesville: University Press of Florida, 2018). Dalton is a languages professor at the University of North Carolina–Charlotte.

52. These sentiments are echoed by other scholars who study technocrats and technology in other parts of Latin America. Examples include Todd A. Diacon, *Stringing Together a Nation: Cândido Mariano da Silva Rondon and the Construction of a Modern Brazil, 1906–1930* (Durham, NC: Duke University Press, 2004); Julia Rodriguez, *Civilizing Argentina: Science, Medicine, and the Modern State* (Chapel Hill: University of North Carolina Press, 2006); María Portuondo, "Constructing a Narrative: The History of Science and Technology in Latin America," *History Compass* 7 (2009): 500–522; Edina Medina, *Cybernetic Revolutionaries: Technology and Politics in Allende's Chile* (Cambridge, MA: MIT Press, 2011); Daniel B. Rood, *The Reinvention of Atlantic Slavery: Technology, Labor, Race, and Capitalism in the Greater Caribbean* (New York: Oxford University Press, 2017); David Pretel, Ian Inkster, and Helge Wendt, eds., "History of Technology in Latin America," special issue of *History of Technology* 34 (2019).

53. Edward Beatty, *Technology and the Search for Progress in Mexico* (Oakland: University of California Press, 2015); Gómez-Galvarriato, *Industry & Revolution*.

54. Some examples include John H. Coatsworth, *Growth against Development: The Economic Impact of Railroads in Porfirian Mexico* (DeKalb: Northern Illinois University Press, 1981); José Antonio Ocampo, *Colombia y la economia nacional, 1830–1910* (Bogotá: Siglo Vientouno de Colombia, 1984); Stephen H. Haber, *Industry and Underdevelopment: The Industrialization of Mexico, 1890–1940* (Stanford, CA: Stanford University Press, 1989); William R. Summerhill, *Order against Progress: Government, Foreign Investment, and Railroads in Brazil, 1854–1914* (Stanford, CA: Stanford University Press, 2003); Sandra Kuntz Ficker, *Las exportaciones mexicanas durante la primera globalización, 1870–1929* (Mexico City: Colegio de México, 2010); Ficker, ed., *La expansión ferroviaria en América Latina* (Mexico City: Colegio de México, 2016).

55. Daniel T. Rodgers, *Atlantic Crossings: Social Politics in a Progressive Age* (Cambridge, MA: Harvard University Press, 1998).

CHAPTER 1: POETRY IN STONE AND IRON

Epigraph: Nicolás Mariscal, "El desarrollo de la arquitectura en Méjico," *El Arte y la Ciencia* 2, no. 10 (January 1900): 1. The translation is mine,

1. Marcela Saldaña Solís, "Luz y espacio: La modernidad en la obra constructi-va de Emilio Dondé Preciat en la ciudad de México," *Boletín de Monumentos Históri-cos* 37 (May–August 2016): 90; Nora Pérez-Rayón Elizundia, *Entre la tradición señorial y la modernidad: La familia Escandón Barrón y Escandón Arango; Formación y desarrollo de la burguesía en México durante el porfirismo (1890–1910)* (Mexico City: UAM-Unidad Azcapotzalco, División de Ciencias Sociales y Humanidades, 1995), 37; and María del Carmen Collado, *La burguesía mexicana: El emporio Braniff y su participación política 1865–1920* (Mexico City: Siglo XXI, 1897), 72.

2. The "lettered city" refers to Angela Rama's *The Lettered City*, ed. and trans. John Charles Chasteen (Durham, NC: Duke University Press, 1996). Rama ex-amines the rise of *letrados*, a Latin American literary elite that became intertwined with state power and elite culture in urban Latin America. For more on class, social status, and the distribution of power, see Max Weber, *The Essential Weber*, ed. Sam Whimster (New York: Routledge, 2004), 182–94.

3. The positivist motto of the Porfirian government.

4. For critical views about adopted new technologies and foreign cultural trends, see Patrick Frank, *Posada's Broadsheets: Mexican Popular Imagery, 1890–1910* (Albuquerque: University of New Mexico Press, 1998); Michael Matthews, *The Civilizing Machine: A Cultural History of Mexican Railroads, 1876–1910* (Lincoln: University of Nebraska Press, 2013).

5. Manuel G Revilla, "Don Lorenzo de la Hidalga," *El Arte y la Ciencia* 3, no. 6 (September 1901): 3.

6. Jaime Rodríguez O, "La Crisis De México en el Siglo XIX," *Estudios de Historia Moderna y Contemporánea de México* 10, no. 10 (1986): 91–92. This second monarchal period in Mexico is often called the Second Empire and the French Intervention.

7. Paul Garner, *Porfirio Díaz: Del héroe al dictador, una biografía política* (Mexico City: Planeta, 2010), 182–83.

8. John Coatsworth, *El impacto económico de los ferrocarriles en el porfiriato* (Mexico City: Ediciones Era, 1976), 43.

9. Vicente Martín Hernández, *La arquitectura doméstica de la ciudad de Méx-ico (1890–1925)* (Mexico City: UNAM, 1981), 25.

10. For more on Porfirian public works, see Priscilla Connolly, *El contratista de Don Porfirio: Obras públicas, deuda y desarrollo* (Mexico City: El Colegio de Micho-acán, 1997); Claudia Agostini, *Monuments of Progress: Modernization and Public Health in Mexico City, 1876–1910* (Calgary: University of Calgary Press, 2003).

11. For more on these topics, see Anne R. Alexander, *City on Fire: Technology, Social Change, and the Hazards of Progress in Mexico City, 1860–1910* (Pittsburgh: University of Pittsburgh Press, 2016); Manuel Perló Cohen, *El paradigma porfiria-no: Historia del desagüe del Valle de México* (Mexico City: Instituto de Investiga-ciones Sociales-UNAM, 1999); Jeffrey M. Banister and Stacie G. Widdifield, "The Debut of 'Modern Water' in Early Twentieth Century Mexico City: The Xochimil-

co Potable Waterworks," *Journal of Historical Geography* 46 (October 2014): 36–52; Matthew Vitz, *A City on a Lake: Urban Political Ecology and the Growth of Mexico City* (Durham, NC: Duke University Press, 2018); J. Justin Castro, *Apostle of Progress: Modesto C. Rolland, Global Progressivism, and the Engineering of Revolutionary Mexico* (Lincoln: University of Nebraska Press, 2019).

12. Manuel Marroquín y Rivera, *Memoria descriptiva de las Obras de Provisión de Aguas Potables para la ciudad de México* (Mexico City: Imprenta y Litografía Müller Hnos, 1914), 3–5; Perló Cohen, *El paradigma porfiriano*, 225.

13. José Valadés, *El porfirismo: Historia de un régimen* (Mexico City: UNAM, 1987), 83–89.

14. Omar Escamilla González, "El laboratorio de resistencia de materiales de construcción de la Escuela Nacional de Ingenieros de México (1892)," *Boletín de Monumentos Históricos* 4 (May–August 2005).

15. Marcela Saldaña Solís, "Luz y espacio: La modernidad en la obra constructiva de Emilio Dondé Preciat en la Ciudad de México," *Boletín de Monumentos Históricos* 37 (May–August 2016): 91–92.

16. The advisers of this professional exam were Vicente Heredia, Juan Cardona, Antonio Torres Torija, and Eleuterio Méndez y Manuel Rincón, important Mexican archetects. See "Acta certificada de Título de Ingeniero y arquitecto Emilio Dondé," 1883/III/221/d.11, Archivo Histórico del Palacio de Minería, Mexico City (hereafter AHPM); Saldaña Solís, "Luz y espacio," 90–91.

17. Leopoldo Rodríguez Morales, *El campo del constructor en el siglo XIX: De la certificación institucional a la esfera pública en la ciudad de México* (Mexico City: Instituto Nacional de Antropología e Historia, 2012), 294–95, 424.

18. Edward Beatty, *Technology and the Search for Progress in Modern Mexico* (Oakland: University of California Press, 2015), 55.

19. Nathanael Grimes and J. Justin Castro, "The Less Imperial Path: The Mississippi Valley, US Expansionism, and Engineer James B. Eads' Failure to Build a Ship Railway," *The Historian* 82, no. 2 (Summer 2020): 156–81.

20. For more about Mexican participation in different world's fairs, see Mauricio Tenorio-Trillo, *Mexico at the World's Fairs: Crafting a Modern Nation* (Berkeley: University of California Press, 1996).

21. Secretaría de Fomento to Emilio Dondé, September 6, 1890, no. 44, book 2, AHPM.

22. Alex M. Saragoza, *The Monterrey Elite & the Mexican State, 1880–1940* (Austin: University of Texas Press, 1988), 55–62.

23. Marcela Saldaña Solís, "Ejemplos y usos del hierro industrial en la obra del ingeniero y arquitecto Emilio Dondé: Ciudad de México (1870–1902)," *Boletín de Monumentos Históricos* 36 (January–April 2016): 104–5.

24. Aurora Gómez-Galvarriato, *Industry & Revolution: Social and Economic Change in the Orizaba Valley, Mexico* (Cambridge, MA: Harvard University Press, 2013), 17.

25. Saldaña, "Luz y espacio," 94.

26. "Rumores y rumorcillos," *La Voz de México*, February 27, 1898, 2.

27. See also Marcela Saldaña Solís, "El Palacio Legislativo Federal y la participación de Emilio Dondé, 1897–1902," *Boletín de Monumentos Histórico* 44 (September–December 2018): 168–79.

28. Secretaría de Comunicaciones y Obras Públicas, "Premios otorgados a los autores del Palacio Legislativo," July 7, 1899, caja 60, exp.530/45, Palacio Legislativo, Archivo General de la Nación, Mexico City.

29. "Concurso para El Palacio Legislativo," *El Tiempo*, April 28, 1898, 4.

30. "Descontento por un fallo," *El Tiempo*, April 26, 1898, 2.

31. Palacio Legislativo, planes 1–158, CNMH-INAH, Archivo Histórico Jorge Enciso, Mexico City (hereafter AHJE).

32. "Palacio Legislativo," *La Voz de México*, July 4, 1901, 2.

33. For more, see Arnaldo Moya Gutiérrez, *Arquitectura, historia y poder bajo el régimen de Porfirio Díaz: Ciudad de México, 1876–1911* (Mexico City: CONACULTA, 2012); Javier Pérez Siller, "México: La nueva traza urbana del poder; Fronteras entre las prácticas porfiristas y su modelo republicano," *Fronteras y sensibilidades en las Américas*, ed. Salvador Bernabéu and Frédérique Langue (Madrid: Ediciones Doce Calles, 2011), 231–58.

34. For more on the subject, see Steven B. Bunker, *Creating Mexican Consumer Culture in the Age of Porfirio Díaz* (Albuquerque: University of New Mexico Press, 2012).

35. The Palacio de Hierro burned down in 1914 and was rebuilt in 1920.

36. "Le premier grand magasin contruir à Mexico," *Le Mexique*, 1904, article qtd. in Gómez-Galvarriato, *Industry & Revolution*, 24.

37. Located at Apartado 233, 1ª Plateros Street, on the corner with Palma Avenue; Donaciones-Emilio Dondé, "Sorpresa y Primavera Unidas (1a. de Plateros)," August, 1893, leg. I, caja 1, plano 15–20, CNMH-INAH, AHJE.

38. J. Figueroa Doménech, *Guía general descriptiva de la República Mexicana* (Mexico City: Ramón de S. N. Araluce, 1899), 256–57.

39. Mónica Silva Contreras, "Arquitectura y materiales modernos: Funciones y técnicas internacionales en la ciudad de México, 1900–1910," *Boletín de Monumentos Históricos* 22 (May–August 2011): 191.

40. "El Lago de Maracaibo: Fábrica de Chocolates," *La Voz de México*, December 1, 1888, 4.

41. From the Sanitary Code, Emilio Dondé cites the following articles: Article 25 Hydraulic lock, Article 9 Common and washing tank; Article 25 Rainwater strainer, Article 10 Fan tube; Article 6 Sewer requirements; Article 21 Grease traps. Planoteca, Puente de Alvarado House no. 11, planes 4 and 5, CNMG-INAH, AHJE.

42. Gómez-Galvarriato, *Industry & Revolution*, 25.

43. "Gacetilla, El Café Colón," *El Tiempo*, July 1, 1891, 2.

44. Agostini, *Monuments of Progress*, 97–100.

45. Donaciones-Emilio Dondé, 2a del Factor Baños, 1881, leg. VI, caja 1, planos 27–28, CNMG-INAH, AHJE.

CHAPTER 2: REVELATIONS FROM REDISCOVERED ARTIFACTS OF THE NATIONAL SCHOOL OF ENGINEERS' CONSTRUCTION MATERIALS COLLECTION

1. Francisco Omar Escamilla González, "El primer laboratorio mexicano de ingeniería civil, hoy Biblioteca 'Ing. Antonio M. Anza,'" in *200 Años del Palacio de Minería: Su historia a partir de fuentes documentales*, ed. Francisco Omar Escamilla González (Mexico City: UNAM-Facultad de Ingeniería, 2013), 364–403.

2. Mireya Blanco y José Omar Moncada Maya, "El Ministerio de Fomento, impulsor del estudio y el reconocimiento del territorio mexicano (1877–1898)," *Investigaciones Geográficas, Boletín del Instituto de Geografía* 74 (April 2011): 74–91.

3. Lucero Morelos Rodríguez, "Historia de las ciencias geológicas: De entidad gubernamental a Instituto Universitario" (PhD diss., Facultad de Filosofía y Letras, UNAM, Mexico City, 2014).

4. For a more detailed account of the origins of engineering and geology education generally in Mexico, see Raúl Domínguez Martínez, *La ingenería civil en México, 1900–1940: Análisis histórico de los factores de su desarrollo* (Mexico City: UNAM, 2013); Mílanda Bazant, "La enseñanza y la práctica de la ingeniería durante el Porfiriato," *Historia Mexicana* 33, no. 1 (1984): 254–97; Lucero Morelos Rodríguez, *La geología mexicana en el siglo XIX: Una revisión histórica de la obra de Antonio del Castillo, Santiago Ramírez y Mariano Bárcena* (Mexico City: Plaza y Valdés-Secretaría de Cultura de Michoacán, 2012), 219–328.

5. Leopoldo Rodríguez Morales, *El campo del constructor en el siglo XIX: De la certificación institucional a la esfera pública en la ciudad de México* (Mexico City: Instituto Nacional de Antropología e Historia, 2012).

6. "Graphic statics is a well-known method for analysis and design of two-dimensional structures based on Cremona's extensions of Maxwell's theory of reciprocal figures. In graphic statics, the relation between form and forces of a structural system is contained in the reciprocal relation between two diagrams. A form diagram describes the geometrical configuration of the (axial) internal and external forces of a two-dimensional structural system, and a force diagram represents their equilibrium. The combination of these two diagrams allows for an intuitive evaluation of structural behavior, performance, and efficiency at a glance. The graphical nature of the method furthermore allows for a visual verification of both the evaluation process and results, making it more transparent than arithmetic or numerical methods"; T. Van Mele and P. Block, "Algebraic Graph Statics," *Computer Aided Design* 53 (2014): 104.

7. Peter Lundgreen, "Engineering Education in Europe and the USA, 1750–1930: The Rise to Dominance of School Culture at the Engineering Professions," *Annals of Science* 47 (1990): 33–75.

8. Lucero Morelos Rodríguez, "Historia de las ciencias geológicas: De entidad gubernamental a Instituto Universitario" (PhD diss., Facultad de Filosofía y Letras, UNAM, Mexico City, 2014).

9. Mónica Silva Contreras, *"Betón Armé* in a Sinking City: Mexico 1902–1914,"* paper presented at the Fourth International Congress on Construction History, Paris, July 2012, 3–7.

10. Mariano Bárcena, *Tratado de Geología: Elementos aplicables a la agricultura, a la ingeniería y a la industria* (Mexico City: Oficina Tip. de la Secretaría de Fomento, 1885), 17. All translations were made by the authors.

11. Bárcena, *Tratado de Geología*, 180, 184–85, 192–93.

12. Lucero Morelos Rodríguez, *La geología mexicana en el siglo XIX: Una revisión histórica de la obra de Antonio del Castillo, Santiago Ramírez y Mariano Bárcena* (Mexico City: Plaza y Valdés-Secretaría de Cultura de Michoacán, 2012), 178.

13. Antonio Torres Torija, *Introducción al estudio de la construcción práctica* (Mexico City: Oficina Tip. de la Secretaría de Fomento, 1885), 144–45.

14. Pablo Argumosa, *Memorándum técnico: Obra compuesta de datos, tablas, y fórmulas útiles para los ingenieros, maestros de obra, industriales, comerciantes, etc.* (Mexico City: Oficina Tip. de la Secretaría de Fomento, 1885), 69.

15. See Alberto Barocio, *Experiencias y estudios, verificados para formular el proyecto de consolidación del subsuelo del Teatro Nacional* (Ciudad Juárez: Edición de Ingeniería, 1923); Ignacio Ulloa del Río, *Palacio de Bellas Artes: Rescate de un sueño* (Mexico City: Universidad Iberoamericana, 2007); Mauricio Tenorio Trillo, *Artilugio de la nación moderna: México en las exposiciones universales, 1880–1900* (Mexico City: Fondo de Cultura Económica, 1998); Daniel Cosío Villegas, *Historia Moderna de México* (Mexico City: Editorial Hermes, 1998).

16. Antonio del Castillo, "Minuta," 1892/II/244/d.4: 2, Archivo Histórico del Palacio de Minería, Mexico City (hereafter AHPM).

17. Antonio del Castillo, *Informe que rinde el director de la Escuela N. de Ingenieros correspondiente al año de 1882* (Mexico City: Oficina Tipográfica de la Secretaría de Fomento, 1884), 41–43.

18. Eduardo Martínez Baca, "Programa para el curso de Conocimiento de Materiales de Construcción y de los terrenos en que deben ejecutarse las obras," 1894/I/248/d.7: 2–4, AHPM. See Enrique Camacho Navarro, "Gilberto Crespo y Martínez, un operador de la diplomacia de México en la Cuba republicana (1902–1906)," *Revista Mexicana de Política Exterior* 84 (July–October 2008): 93–120; Enrique Camacho Navarro, "Gilberto Crespo y Martínez y su participación en la política de fomento para el México Porfirista: Reflexiones a propósito de su obra dedicada a Bélgica," *Tzintzun. Revista de Estudios Históricos* 49 (January–July 2009): 131–68.

19. Martínez Baca, "Programa," 2–4.

20. Felix Karrer, *Führer durch die Baumaterial-Sammlung des k.k. naturhis-*

torischen Hofmuseum in Wien (Viena: Verlag für R. Lechner's k. u. k. Hof- und Univ.-Buchhandlung, 1892).

21. *Catalogue des échantillons de matériaux de construction réunis par les soins du Ministère des Travaux Publics* (Paris: Dunod, 1878).

22. This fits with a broader trend in which Mexican diplomats were regularly sending information about trends and new technologies from Europe. For example, historian Justin Castro points out that Mexican diplomats in Europe in the late 1890s and early 1900s were crucial to the spread of early radiotelegraph technologies and ideas about how they could be used for nation-state consolidation. See J. Justin Castro, *Radio in Revolution: Wireless Technology and State Power, 1897–1938* (Lincoln: University of Nebraska Press, 2016), chap. 1.

23. Antonio del Castillo, "Minuta," 1882/I/215/d. 39: 3–4, AHPM,.

24. Del Castillo, "Minuta," 1882, 1.

25. Del Castillo, "Minuta," 1882, 1–1v.

26. "Anexo 9: Circular para la formación de colecciones de materiales de construcción," in *Memoria presentada al Congreso de la Unión por el Secretario de Estado y del Despacho de Fomento, Colonización, Industria y Comercio de la República Mexicana, corresponde a los años transcurridos de enero de 1883 a junio de 1885* (Mexico City: Oficina Tip. de la Secretaría de Fomento, 1887), 4:506–7.

27. Antonio del Castillo, qtd. in Carlos Pacheco, "Minuta," 1882/II/216/d. 22, AHPM.

28. Manuel Fernández Leal, "Oficio," 1882/II/216/d. 67, AHPM. It is possible that some of the existing samples in Palacio de Minería could be part of this shipment since the handwriting, ink, and labeling were probably made at that time. Label information on the samples reads: "No. 1. Cantera. Estado de Guanajuato. Hacienda de Cieneguita. Para columnas, cornisas y fuentes" (Quarry, State of Guanajuato, Hacienda of Cieneguita. For columns, cornices and fountains); "No. 2. Ripio. Allende (Estado de Guanajuato) Valle del maíz. Para construcciones urbanas" (Gravel. Allende [State of Guanajuato] Maiz Valley. For urban constructions); "No. 4. Piedra de Cerro. Estado de Guanajuato. Cuesta de Rondanejo. Para calzadas y muros" (Hill Stone. State of Guanajuato, Rondanejo Hill. For roads and walls); "No. 6. Cantera. Estado de Guanajuato, Hacienda de la Virgen. Para columnas y cornisas" (Quarry, State of Guanajuato. Hacienda of La Virgen. For columns and cornices).

29. Santiago Ramírez, "Noticia de las rocas y materiales de construcción recogidos en Ixtapan de la Sal y remitidos a la Secretaría de Fomento," 1882/IV/218/d. 52: 3v, AHPM.

30. Ramírez, "Noticia," 3v.

31. Ramírez, "Noticia," 3v–4.

32. Ramírez, "Noticia," 4v.

33. Ramírez, "Noticia," 4v.

34. Ramírez, "Noticia," 6v.

35. Ramírez, "Noticia," 5v.

36. Ramírez, "Noticia," 5v–6.

37. Adrián Aguilar Contreras, *Reporte sobre los barrios e industria alfarera que los utiliza, del pueblo de Tecomatepec, munic. de Ixtapan de la Sal, Méx.* (Mexico City: Servicio Geológico Mexicano, 1970), https://mapserver.sgm.gob.mx/informes/textos/T1570AUCA0034_01.pdf.

38. Torres Torija, *Introducción*, 27–29, 37–40.

39. Ramírez, "Noticia," 7v.

40. Torres Torija, *Introducción*, 29; Argumosa, *Memorándum*, 75.

41. Torres Torija, *Introducción*, 40. One carretada equals 10 cargas of 12 arrobas each, adding up to 120 arrobas or 1,380 kg (1 arroba = 11.5 kg).

42. Argumosa, *Memorándum*, 74.

43. Rubén Antonio Vega González and Roberto Reyes Pérez, "La arquitectura de madera en el Porfiriato yucateco," paper presented at the Segundo Coloquio Mexicano de Historia de la Construcción, Mérida, October, 2016.

44. Héctor Vera, *A peso el kilo: Historia del sistema métrico decimal en México* (Mexico City: Libros del Escarabajo, 2007).

45. Ramírez, "Noticia," 5v.

46. Torres Torija, *Introducción*, 29.

47. Aguilar Contreras, *Reporte*; Adrián Aguilar Contreras, *Reporte sobre la pizarra arcillosa que ocurre cerca de la ciudad de Ixtapan de la Sal, Méx, y la cual se estima es útil para la manufactura de tabiques* (Mexico City: Servicio Geológico Mexicano, 1970),

https://mapserver.sgm.gob.mx/informes/textos/T1570AUCA0030_01.pdf.

48. See Andrés Molina Enríquez, *Los grandes problemas nacionales* (Mexico City: Secretaría de Cultura, Instituto Nacional de Estudios Históricos de las Revoluciones de México, 2016).

49. *Fundadores del Instituto de Ingeniería: Inteligencia y pasión* (Mexico City: UNAM-Instituto de Ingeniería, 2014).

CHAPTER 3: THE PREOCCUPATION WITH SAFETY

Portions of this chapter are reproduced from Rocio Gomez, *Silver Veins, Dusty Lungs: Mining, Water, and Public Health in Zacatecas, 1835–1946* (Lincoln: University of Nebraska Press, 2020), with permission of the University of Nebraska Press. Copyright 2020 by the Board of Regents on the University of Nebraska. All Spanish translations are mine unless otherwise noted.

1. Gregorius Agricola, *De re metallica*, trans. Herbert C. Hoover and Lou Henry Hoover (New York: Dover, 1950), 214–16; originally published in 1912 by *Mining Magazine*, London.

2. Bernardino Ramazzini, *De morbis artificum diatriba (Diseases of Workers)*, trans. Wilmer Cave Wright (1700; Chicago: University of Chicago Press, 1940), 2.

3. Ramazzini, *De morbis artificum diatriba*, 2. For comparative histories on occupational health and safety, see Martin Cherniack, *The Hawk's Nest Incident: America's Worst Industrial Disaster* (New Haven, CT: Yale University Press, 1989); Claudia Clark, *Radium Girls: Women and Industrial Health Reform* (Chapel Hill: University of North Carolina Press, 1997); Kate Brown, *Plutopia: Nuclear Families, Atomic Cities, and the Great Soviet and American Plutonium Disasters* (New York: Oxford University Press, 2015); Kate Moore, *The Radium Girls: The Dark Story of America's Shining Women* (New York: Sourcebooks, 2017); Ángela Vergara, *Fighting Unemployment in Twentieth-Century Chile* (Pittsburgh: University of Pittsburgh Press, 2021); David M. Turner and Daniel Blackie, *Disability in the Industrial Revolution: Physical Impairment in British Coalmining, 1780–1880* (Manchester, UK: Manchester University Press, 2018); David Rosner and Gerald Markowitz, *Deadly Dust: Silicosis and the Politics of Occupational Disease in Twentieth-Century America* (Lansing: University of Michigan Press, 1991). For the intersection of labor and the environment, see Myrna I. Santiago, *The Ecology of Oil: Environment, Labor, and the Mexican Revolution, 1900–1938* (New York: Cambridge University Press, 2006); Gomez, *Silver Veins, Dusty Lungs*; and Nicholas A. Robins, *Mercury, Mining, and Empire: The Human and Ecological Cost of Silver Mining in the Andes* (Bloomington: Indiana University Press, 2011).

4. Sara B. Pritchard and Carl A. Zimring, *Technology and the Environment* (Baltimore: Johns Hopkins University Press, 2020), 109–10.

5. See Turner and Blackie, *Disability in the Industrial Revolution.*

6. See Alan Derickson, *Black Lung: Anatomy of a Public Health Disaster* (Ithaca, NY: Cornell University Press, 2011).

7. See Joch McCulloch, *Asbestos Blues: Labour, Capital, Physicians, and the State of South Africa* (Bloomington: Indiana University Press, 2002); Joch McCulloch, *Defending the Indefensible: The Global Asbestos Industry and Its Fight for Survival* (New York: Oxford University Press, 2008).

8. Ángela Vergara, "The Recognition of Silicosis: Labor Unions and Physicians in the Chilean Copper Industry, 1930s–1960s," *Bulletin of the History of Medicine* 79, no. 4 (Winter 2005): 728–35. See also Vergara, *Fighting Unemployment.*

9. Historical works on colonial mining in Mexico include Danna Velasco Murillo, "Borderlands in the Silver Mines of New Spain, 1540–1660," in *Oxford Handbook of Borderlands*, ed. Danna A. Levin Rojo and Cynthia Radding (New York: Oxford University Press, 2019), 371–95; Kendall W. Brown, *A History of Mining in Latin America* (Albuquerque: University of New Mexico Press, 2012); P. J. Bakewell, *Silver Mining and Society in Colonial Mexico, Zacatecas 1546–1700* (Cambridge: Cambridge University Press, 1971); David A. Brading, *Miners, and Merchants in Bourbon Mexico, 1763–1810* (Cambridge: Cambridge University Press, 1971).

10. Friedirch Traugott Sonneschmidt, *Tratado de la Amalgamación de México*

(Mexico City: Redacción de "El Minero Mexicano," 1876), 95–96; originally published by Imprenta de D. Mariano de Zúñiga y Ontiveros in 1805.

11. Allison Margaret Bigelow, *Mining Language: Racial Thinking, Indigenous Knowledge and Colonial Metallurgy in the Early Modern Iberian World* (Chapel Hill: University of North Carolina Press, 2020), 251.

12. Santiago Ramírez, *Datos para la historia del Colegio de Minería*, edición de la Sociedad "Antonio Alzate" (Mexico City: Exprenta del Gobierno Federal en el Ex-Arzobispado, 1890), 20–21.

13. Juan Lucas de Lessage and Joaquín Velázquez de León, qtd. in Ramírez, *Datos para la historia*, 20–21. The mining school model had already taken root in Spain, according to María Dolores Muñoz Dueñas, "La formación de una élite minera: La Escuela de Minas de España (1777–1877)," *Mélanges de la Casa de Velásquez* 28, no. 3 (1992): 21–36.

14. See Bigelow, *Mining Language*; and Omar Escamilla González and Lucero Morelos Rodríguez, *Escuelas de Minas Mexicanas: 225 Años del Real Seminario de Minería, México* (Mexico City: Universidad Nacional Autónoma de México, 2017), for an analysis of early modern introductions of European scientific concepts regarding mining amalgamation. For primary sources, see Ramírez, *Datos para la historia*.

15. Missive from El Ministro de Indias to Virrey de Nueva España, April 19, 1798, repr. in Ramírez, *Datos para la historia*, 143.

16. Until the War of Independence (1810–1821), Fausto de Elhuyar attempted to install mining education centers in traditional mining cities in New Spain. Notably, in Guanajuato he hired faculty in 1797 for chemistry, mathematics, and physics at the Real Colegio de la Purísima Concepción, originally established in 1785.

17. Ramírez, *Datos para la historia*, 366.

18. Editorial, *El Eco Nacional*, February 18, 1857, repr. in Ramírez, *Datos para la historia*, 397–98.

19. See Escamilla González and Morelos Rodríguez, *Escuelas de Minas Mexicanas*.

20. "La Educación Práctica de los Ingenieros de Minas de México," *Boletín Oficial de la Cámara Minera de México* 9 (February 1911): 2; "La Educación Práctica de los Ingenieros de Minas de México," *Boletín Oficial de la Cámara Minera de México* 10 (March 1911): 2.

21. "Intervención del Gobierno en los Trabajo de las Minas," *Boletín Oficial de la Cámara Minera de México* 7 (December 1910): 2.

22. "La Educación Práctica de los Ingenieros de Minas de México," *Boletín Oficial de la Cámara Minera de México* 11 (April 1911): 1.

23. "La Educación Práctica," April 1911, 1.

24. "La Educación Práctica," April 1911, 2.

25. See Gerardo Tanamachi Castro and María de la Paz Ramos Lara, "La

escuela nacional de ingenieros y las ciencias físicas en los albores del Siglo XX," *Revista Mexicana de Investigación Educativa* 20, no. 65 (2015): 557–80.

26. Escamilla González and Morelos Rodríguez describe the development of the curriculum during the colonial period in *Escuelas de Minas Mexicanas.*

27. "La Educación Práctica de los Ingenieros de Minas de México," *Boletín Oficial de la Cámara Minera de México* 14 (July 1911): 2.

28. "La Educación Práctica," July 1911, 2.

29. "Intervención del Gobierno," *Boletín Oficial de la Cámara Minera de México* 7 (December 1910): 1.

30. "Intervención del Gobierno," 1–2.

31. "Intervención del Gobierno," 2.

32. "Intervención del Gobierno," 2.

33. "Intervención del Gobierno," 2–3.

34. Código de Minería 1884, Título VI, Art. 119–25.

35. Código de Minería 1892, Título I, Art. 12, sections XII–XIV.

36. Mexican Constitution of 1917, Article 123, section XIV.

37. Mexican Constitution of 1917, Article 123, section XIV.

38. "Circular sobre dinamita . . . ," Jefatura Política, Caja 2 (1880–1889), Exp. N/A, Archivo Histórico del Estado de Zacatecas, Zacatecas, Mexico (hereafter AHEZ).

39. "Juan Petit pide permiso . . . ," Ayuntamiento, Exp. N/A, AHEZ.

40. "Aviso sobre robo de dinamita . . . ," Ayuntamiento, Minería, Caja 2 (1823–1889), Exp. N/A, AHEZ.

41. "Intervención del Gobierno," 2.

42. Luis Canales, judicial hearing testimony, April 28, 1891, Poder Judicial, Criminal, Caja 6, Exp. 150, AHEZ.

43. Francisco Zarate and Luis Cordova, judicial report, January 18, 1891, Poder Judicial, Criminal, Caja 6, Exp. 150, AHEZ.

44. Enrique Carrillo to Jefatura Política, October 3, 1884, Jefatura Política, Serie Correspondencia General, Caja 1 (1823–1899), Exp. N/A, AHEZ.

45. See Anna R. Alexander, *City of Fire: Technology, Social Change, and the Hazards of Progress in Mexico City, 1860–1910* (Pittsburgh: University of Pittsburgh Press, 2016); and J. Justin Castro, *Apostle of Progress: Modesto C. Rolland, Global Progressivism, and the Engineering of Revolutionary Mexico* (Lincoln: University of Nebraska Press, 2019), for examples of Mexican engineers' growing sense of social responsibility in fire prevention and infrastructure, respectively, during this era.

46. Lewis Mumford, *Technics and Civilization* (New York: Harcourt Brace, 1934; Chicago: University of Chicago Press, 2010), 10.

47. Judge F. Henríquez, judicial report, February 13, 1891, Poder Judicial, Serie Criminal, Caja 21, Exp. 537, AHEZ.

48. Judge F. Henríquez, judicial report, February 27, 1891, Poder Judicial, Serie Criminal, Caja 21, Exp. 537, AHEZ.

49. Crecensio Ríos, testimony before Judge Jesús Zamora, October 31, 1891, Poder Judicial, Serie Criminal, Caja 22, Exp. 575, AHEZ.

50. Crecensio Ríos, testimony before Judge Jesús Zamora, October 31, 1891, Poder Judicial, Serie Criminal, Caja 22, Exp. 575, AHEZ.

51. Judge Jesús Zamora, judicial report, October 31, 1891, Poder Judicial, Serie Criminal, Caja 22, Exp. 575, AHEZ.

CHAPTER 4: ENGINEERING THE PORFIRIAN LANDSCAPE

1. José Ramón de Ibarrola, *Apuntes sobre el desarollo sobre el ingeniería en México y le educación en ingenierio* (Mexico City: Tipografia de la F. Diaz de Leon, 1911). The academy was founded in 1890 with the intent to centralize and disseminate knowledge from Mexico's legal, scientific, and art academies; see Alejandro Mayagoitia, "El Concurso cientifico y artistico del centenario de la independencia o la historia del derecho como ditirambo," *Anuario Mexicano de Historia del Derecho* 13 (2001): 30–40.

2. For a history of the desagüe during the colonial era, see Vera S. Candiani, *Dreaming of Dry Land: Environmental Transformation in Colonial Mexico City* (Stanford: Stanford University Press, 2014).

3. Edward Beatty, *Technology and the Search for Progress in Modern Mexico* (Oakland: University of California Press, 2015), 1–7.

4. Priscilla Connolly, *El contratista de don Porfirio: Obras publicas, deuda y desarollo desigual* (MexicoCity : Fondo de Cultura Economica, 1997), 300–304.

5. Raúl Domínguez Martínez, *La ingeniería civil en México, 1900–1940: Análisis Histórico de los factores de su desarrollo* (Mexico City: Universidad Nacional Autónoma de México, 2013), 278.

6. Domínguez Martínez, *La ingeniería civil*, 74–76. In addition to Connolly, who devotes several chapters to the Gran Canal del Desagüe's financial aspects, the most complete political and economic coverage is Manuel Perló Cohen, *El paradigma porfiriano: Historia del desagüe del Valle de México* (Mexico City: Porrúa, 1999), while Paul Garner's *British Lions and Mexican Eagles: Business, Politics and Empire in the Career of Weetman Pearson in Mexico 1889–1919* (Stanford, CA: Stanford University Press, 2011) examines the career of the head of S. Pearson and Son in the context of Mexican politics and the firm's projects in Mexico, including the Gran Canal.

7. Market development in the greater region had long been tied to silver mining and had seen a renewed boom since independence. See Rodolfo Ramírez Rodríguez, "La formación de un mercado regional en el noreste del valle de México: De minas, ferrocarril y haciendas pulqueras (1850–1870)," *Anuario de Historia Regional y de las Fronteras* 22, no. 1 (2017): 17–48.

8. See Vera Candiani, "The Desagüe Reconsidered: Environmental Dimensions of Class Conflict in Colonial Mexico," *Hispanic American Historical Review* 92, no. 1 (2012): 10–13.

9. Garner, *British Lions and Mexican Eagles*, 64–70.

10. Luis Gonzalez Obregón, report in *Memoria histórica, técnica y administrativa de las obras del desagüe del valle de México, 1449–1900* (Mexico City: Tipografía de la Oficina Impresora de Estampillas, 1902), 1:ix.

11. De Garay had worked on numerous projects in Mexico and was thoroughly familiar with the hydrology of the valley. See Francisco de Garay, *El Valle de México: Apuntos históricos sobre su hidrográfica* (Mexico City: Oficina Tipografía de la Secretaría de Fomento, 1888); Gonzalez Obregón, *Memoria*, viii.

12. *Drainage Works in the Valley of Mexico: Information written for the Members of the Society of American Civil Engineers* (Mexico City: Tipografía de la Direccion General de Telegrafos, 1907), 20.

13. Allen Boyer McDaniel, *Excavating Machinery* (New York: McGraw Hill Book Company, 1913), 198–216.

14. *Drainage Works in the Valley of Mexico*, 26.

15. J. A. Spender, *Weetman Pearson: First Viscount Cowdray, 1856–1927* (London: Cassell, 1930), 93.

16. Luis Espinosa, "Reseña Técnica de la ejecución del gran canal y de las obras del arte 1886–1900," in *La memoria histórica, técnica y administrativa*, 1:499–508.

17. Thomas D. Rogers, *The Deepest Wounds: A Labor and Environmental History of Sugar in Northeast Brazil* (Chapel Hill: University of North Carolina Press, 2010), 6.

18. Charles Gibson, *Aztecs under Spanish Rule* (Stanford, CA: Stanford University Press, 1964), 338–44; Ursula Ewald, *The Mexican Salt Industry, 1560–1980* (New York: G. Fischer, 1985), 30–41.

19. Francisco de Garay, *El valle de México: Apuntos historicos sobre su hidrografia, desde los tiempos mas remotos hasta nuestros dias* (Mexico City: Oficina Tipografía de la Secretaria de Fomento, 1888), 91–92.

20. See Antonio Peñafiel and Lamberto Asiain, *Memoria sobre las Aguas Potables de la Capital de México* (Mexico City: Secretaría de Fomento, 1884), 127–32 (my translation).

21. Rosendo Esparza, "Reseña administrativa y económica de la Junta Directiva del Desagüe del Valle de México, 1886–1900," in *La memoria histórica, técnica y administrativa*, 1:607.

22. Jefe Politico Zumpango, to Junta Directiva del Desagüe, February 28, 1891, Junta Directiva de los Trabajos del Desagüe del Valle de Mexico, Exp. 240, *Sobre quejas y peticiones diversas de los pueblos por donde pasa el Gran Canal*, Caja 18, FO180, Serie Desagüe del Valle de México, Comunicaciones y Obras Publicas, (hereafter Exp. 240, quejas y peticiones, Serie del Desagüe, Archivo General de la Nación, Mexico City (hereafter AGN).

23. Antonio García Cubas, *Diccionario geográfico, histórico y biográfico de los Estados Unidos Mexicanos* (Mexico City: Oficina Tipográfica de la Secretaría de Fomento, 1898), 5:521–23.

24. Luis Espinosa to Junta Directiva del Desagüe, March 14, 1891, Junta Directiva de los Trabajos del Desagüe del Valle De Mexico, Exp. 240, quejas y peticiones, Serie del Desagüe, AGN.

25. Some of these problems included wear and tear on dredge scoops, disagreements over how much the junta should pay Pearson for the labor cost of the material each dredge scooped out, and delays on the project's completion. See Cohen, *El paradigma Porfiriano*, 159–74.

26. Peticion to Junta Directiva, September 13, 1892, Junta Directiva de los Trabajos del Desagüe del Valle De Mexico, Exp. 240, quejas y peticiones, Serie del Desagüe, AGN.

27. Manuel Orozco y Berra, *Memoria para la carta hidrográfica del Valle de Mexico* (Mexico City: Imprenta de A. Boix, 1864), 161.

28. Romana Falcón, "El arte de la petición: Rituales de obediencia y negociación, Mexico, segunda mitad del siglo XIX," *Hispanic American Historical Review* 86, no. 3 (August 2006): 482–83.

29. Gabino Pineda to Junta Directiva del Desagüe, August 21, 1891, Junta Directiva de los Trabajos del Desagüe del Valle de Mexico, Exp. 240, quejas y peticiones, Serie del Desagüe, AGN.

30. Isidro Díaz Lombardo to Junta Directiva de los Trabajos del Desagüe del Valle de Mexico, December 16, 1891, Exp. 240, quejas y peticiones, Serie del Desagüe, AGN.

31. See Raymond B. Craib, *Cartographic Mexico: A History of State Fixations and Fugitive Landscapes* (Durham, NC: Duke University Press, 2004), 198–99.

32. Residents of San Juan de Aragón and other pueblos in the area would continue to battle for their rights to access land and resources after the end of the Porfiriato. See Matthew Vitz, *A City on a Lake: Urban Political Ecology and the Growth of Mexico City* (Durham, NC: Duke University Press, 2018).

33. Perlo Cohen, *El paradigma Porfiriano*, 161–247.

34. Jose Mena y Díaz to Junta Directiva de los Trabajos del Desagüe del Valle de Mexico, October 10, 1894, Exp. 240, quejas y peticiones, Serie del Desagüe, AGN.

35. Patricia Romero Lankao, *Obra hidráulica de la ciudad de México y su impacto socioambiental, 1880–1990* (Mexico City: Instituto Mora, 1990), 74.

36. "Sistema de riego del Valle de Mezquital," *Irrigación en Mexico* 1, no. 2 (May 1930): 11–21.

37. Connolly, *El contratista*, 304.

38. Domínguez Martínez, *La ingeniería civil*, 84.

39. See Garner, *British Lions and Mexican Eagles*.

40. *Breve reseña de las obras del desagüe del valle de México: Escrita expresamente para los delegados al Congreso pan-americano.* (Mexico City: Tipografía de F. Díaz de Leon, 1901).

CHAPTER 5: REVOLUTIONARY TECHNOSCIENCE

1. *Actas y Memorias del Primer Congreso Científico Mexicano* (Mexico City: Imprenta del Museo Nacional de Arqueología, Historia y Etnología, 1913), 45.

2. *Crónicas y debates de las sesiones de la Soberana Convención Revolucionaria*, vol. 1, Sesión del día 3 de octubre de 1914 (Mexico City: INEHRM, 1964), 55.

3. For the political, scientific, and technological history of the revolution, see Juan José Saldaña, *Las revoluciones políticas y la ciencia en México*, vol. 2 (Mexico City, CONACYT, 2010), chaps. 3–5.

4. The engineer Félix Palavicini, while in charge of the Office of the Ministry of Public Instruction and Fine Arts, presented the new administrative plan in educational instruction and the political criteria on which it was based, giving possession to engineer Valentín Gama, the new rector of the university. See *Boletín de Educación* 1, no. 1 (September 1914).

5. Among others, in 1915 the following engineers were collaborating with the provisional government: Pastor Rouaix, Alberto J. Pani, and Félix Palavicini (in charge of the Ministries of Development, Industry, and Education, respectively); the naturalist Alfonso L. Herrera (Director of the Bacteriologist Institute); the petroleum geologist Ezequiel Ordoñez (Director of the Geological Institute); the chemist Salvador Agraz (director of the Industrial Chemical School); Dr. Guadalupe Gracia (director of the Military Medical School); the mathematician Pedro C. Sánchez (director of Geographic and Climatologic Studies); and the numerous engineers who took part in the diverse technical commissions that were formed during this time, such as Modesto C. Rolland and José Vázquez Schiaffino, who were members of the Petroleum Technical Commission formed in the same year.

6. Venustiano Carranza, qtd. in "Al abrir las sesiones extraordinarias del Congreso, el 15 de abril de 1917," in *Los Presidentes de México ante la Nación* (Mexico City: Cámara de Diputados, 1966), 3:164. This and all other translations not otherwise attributed are my own. For the important regulatory and technical activities of the Technical Petroleum Commission created in 1915, see Saldaña, *Las revoluciones políticas y la ciencia*, 2:240; and J. Justin Castro, *Apostle of Progress: Modesto C. Rolland, Global Progressivism, and the Engineering of Revolutionary Mexico* (Lincoln: University of Nebraska Press, 2019), 58–65.

7. See *Informe presentado a la Secretaría de Comunicación Obras Públicas por el Gerente General Int. Teniente Coronel Ingeniero J. Morales Hesse* (Mexico City: Talleres Tipográficos de la Compañía de Tranvías de México, 1915). For the important case of the domiciliation of technology, see Haydé Toledo and Juan José Saldaña, "La tecnología de la gestión en la Compañía de Tranvías de México entre el Porfiriato y la Revolución," in *Conocimiento y acción: Relaciones históricas de la ciencia, la tecnología y la sociedad en México*, ed. Juan José Saldaña (Mexico City: Facultad de Filosofía y Letras-UNAM y Plaza y Valdés, 2013), 207–31.

8. "Se creará dentro de poco el Departamento de Avicultura," *El Demócrata*, March 6, 1917, 5.

9. See Juan Manuel Cervantes Sánchez and Juan José Saldaña, "Las estaciones agrícolas experimentales en México (1908–1921) y su contribución a la ciencia agropecuaria mexicana," in *La Casa de Salomón en México: Estudios sobre la institucionalización de la docencia y la investigación científicos*, ed. Saldaña (Mexico City: Facultad de Filosofía y Letras, UNAM, 2005), 334.

10. David Baptista and Juan José Saldaña, "La participación política y reivindicación gremial del Centro de Ingenieros de México ante la construcción del Estado Mexicano en los años veinte," in *Memorias del Primer Coloquio Latinoamericano de Historia y Estudios Sociales sobre la Ciencia y la Tecnología*, ed. Guadalupe Urbán, CD-ROM (Mexico City: Sociedad Mexicana de Historia de la Ciencia y la Tecnología, 2007): 1228–30. The Center for Engineering was an important and influential organization that needs more thorough study.

11. On January 8, 1921, Rector Vasconcelos wrote to the center, "Desiring to proceed with the reorganization of the National Schools of Engineers, the undersigned Rector has believed it advisable to consult that honorable association about plans, programs and personnel, and in this manner, I sincerely ask you, if you do not find it inconvenient, to utilize the commission to undertake the case studies as soon as possible since the courses must be open on February 1 and remit your opinion to this office under my charge"; repr. in "La reorganización de la Escuela Nacional de Ingenieros," *Boletín de la Universidad* 2, no. 4 (March 1921): 175.

12. "Puntos resolutivos del Centro de Ingenieros sobre la consulta del C. Rector de la Universidad Nacional acerca de la Escuela Nacional de Ingenieros," *Boletín de la Universidad* 2, no. 4 (March 1921): 176–86.

13. "Puntos resolutivos," 176–86.

14. Daniel Reséndiz Núñez, *La investigación en ingeniería: Consideraciones sobre su historia en México* (Mexico City: Instituto de Ingeniería, UNAM, 1979), 7.

15. For more on these ideas, see the Mexico City newspaper *El Universal*, December 3, 1925, and December 10, 1925, front page.

16. During this time the Mexican government began to utilize its own oil wells, assigned to the railways and the Administration of Petroleum Office, antecedents of PETROMEX, the latter formed in 1934; the distribution of oil from these wells was assigned to public works. See Reyes Edgar Castañeda Crisolis, "Los desafíos técnicos y tecnológicos de la Expropiación Petrolera en México: El papel del Estado y la comunidad científica y tecnológica" (PhD diss., Facultad de Filosofía y Letras, UNAM, Mexico City, 2011).

17. Ángel Peimbert, "Velada conmemorativa del 1er. Centenario del nacimiento del Sr. Ing. D. Manuel M. Contreras," *Revista Mexicana de Ingeniería y Arquitectura* 12, no. 1 (January 1934): 18. The historian Enrique Cárdenas also cites "a rapid increase in value of exports, mainly in silver and oil, and a series

of expansionist monetary and fiscal policies" for the economic recovery; Cárdenas, "La Gran Depresión y la industrialización: El caso de México," in *Historia económica de México*, ed. Cárdenas (Mexico City: Fondo de Cultura Económica, 1994), 5:22.

18. "Plan Sexenal 1934–1940," in *Antología de la Planeación en México, 1917–1985* (Mexico City: Secretaría de Programación y Presupuesto-Fondo de Cultura Económica, 1985), 1:191. On the authorship of the idea of government planning, other sources attribute it to Secretary of Education Narciso Bassols, whose socialist positions were more radical, and therefore the text of the plan reveals a lack of theoretical uniformity. See Victoria Lerner, "El Plan Sexenal de 1933," *Estudios políticos*, nueva época 6, nos. 1 and 2 (1975): 23.

19. Lerner, "El Plan Sexenal de 1933," 192.

20. Lerner, "El Plan Sexenal de 1933," 201.

21. Lerner, "El Plan Sexenal de 1933," 209.

22. Lerner, "El Plan Sexenal de 1933," 213.

23. Enrique Beltrán, *Medio siglo de recuerdos de un biólogo mexicano* (Mexico City: Sociedad Mexicana de Historia Natural, 1977), 115.

24. Enrique Beltrán, "Campos de actividad en los que preferentemente debe desarrollar sus labores el Instituto Biotécnico, para que éstas tengan una repercusión social revolucionaria," typewritten document, December 3, 1933, 8 pp., Beltrán personal archive, Centro de Estudios Dr. Enrique Beltrán, Instituto Mexicano de Recursos Naturales Renovables, Mexico City (hereafter CEDEB),.

25. Beltrán, "Campos de actividad."

26. Enrique Beltrán, "Instituto Biotécnico," *Anales de la Sociedad Mexicana de Historia de la Ciencia y de la Tecnología*, no. 1 (1969): 174.

27. Lázaro Cárdenas to Ing. Antonio Rojas Garcia, telegram, May 24, 1934, CEDEB.

28. Enrique Beltrán, *Lista de peces mexicanos*, mimeo (Mexico City: Instituto Biotécnico,1934).

29. "Opinion of Dr. José G. Parres, Secretary of Agriculture, on the Reorganization of the Biotechnical Institute," December 10, 1939, CEDEB.

30. "El Instituto de Salubridad y Enfermedades Tropicales y sus funciones," *Revista del Instituto de Salubridad y Enfermedades Tropicales* 1 (1939–1940): 6.

31. See José Álvarez Amézquita, Miguel Bustamante, et al., *Historia de la Salubridad y la Asistencia en México*, vol. 2 (Mexico City: Secretaría de Salubridad y Asistencia, 1960).

32. See *Organización, Planes y Programas de Estudios de la Facultad de Ciencias Físicas y Matemáticas de la Universidad Nacional de México (1935)*, facsimile ed. (Mexico City: Facultad de Ciencias-Prensas de Ciencias, 1991).

33. For the foundation and the initial activities of Instituto Politécnico Nacional, see Jesús Ávila Galinzoga, ed., *La educación técnica en México desde la Independencia, 1810–2010*, vol. 2 (Mexico City: Instituto Politécnico Nacional, 2011).

34. The National Commission of Irrigation did still rely on US engineering advice and superintendents for certain major dam projects from 1936 to 1946; see Mikael D. Wolfe, *Watering the Revolution: An Environmental and Technological History of Agrarian Reform in Mexico* (Durham, NC: Duke University Press, 2017).

35. *La obra de la Comisión Nacional de Irrigación durante el régimen del Sr. General de División Lázaro Cárdenas. 1934–1940* (Mexico City, 1940).

36. *La obra de la Comisión Nacional de Irrigación*, 118.

37. Luis Aboites and Engracia Loyo, "La construcción del nuevo Estado, 1920–1943," in *Nueva Historia General de México*, ed. Erick Velázquez García et al. (Mexico City: El Colegio de México, 2010), 632.

38. Cárdenas, "La Gran Depresión y la industrialización," 23.

39. See Stephen H. Haber, "Recuperación y crecimiento, 1933–1940," in Cárdenas, *Historia económica de México*, 298.

40. See Cárdenas, introduction to *Historia económica de México*, 27.

41. For a view of the convergence of the development of the engineering trades in Mexico, see Óscar M. González Cuevas, "Ingeniería, Diseños y Tecnología," in *Cosmos: Enciclopedia de las ciencias y la tecnología en México*, ed. Carlos Herrero, vol. 1 (Mexico City: CONACYT-UAM- ICYTDF, 2010).

42. The PRM was the immediate predecessor to the PRI.

43. "Segundo Plan Sexenal 1940–1946," in *Antología de la Planeación en México (1917–1985)* (Mexico City: Fondo de Cultura Económica, 1985), 1:274.

44. "Discurso pronunciado, el dia 3 de Noviembre de 1939, por el C. General Manuel Ávila Camacho, al otorgar su protesta la honorable asamblea nacional del Partido de la Revolución Mexicana, como candidato a la Presidencia de la República," in *Antología de la Planeación en México*, 1:338–39.

45. The Central de Trabajadores de Mexico prepared the Second Plan Sexenal in February 1939 and in November of that year a commission designated by the PRM's congress elaborated a final version containing the guidelines for moderation and inclusion that had resulted from political negotiations among the party's sectors, which were to guide the campaign and government of Ávila Camacho. On Ávila Camacho's candidacy and the second Sexenal, see Luis Medina Peña, *Del cardenismo al avilacamachismo*, vol. 18, *Historia de la Revolución Mexicana* (Mexico City: El Colegio de México, 1978), 85.

46. "Segundo Plan Sexenal," 284.

47. "Segundo Plan Sexenal," 287.

48. "Segundo Plan Sexenal," 293.

49. "Segundo Plan Sexenal," 312.

50. "Segundo Plan Sexenal," 305.

51. For the characteristics that Mexican industry adopted from the Porfiriato until the end of Cardenismo and industrial modernization, see Stephen Harber, *Industria y subdesarrollo: La industrialización de México, 1890–1940* (Mexico City: Editorial Alianza, 1992).

52. Industrial development would be oriented toward complex industries such as cotton textiles, wool, rayon, chemical, cement, paper, food, sugar, the production of agricultural machinery, electrical equipment, ceramics, aluminum, and other industrial projects, which made this task unfeasible for a single government. See Stanford A. Mosk, *Industrial Revolution in Mexico* (Berkeley: University of California Press, 1954).

53. Artículo 12° de la "Ley de Industrias de Transformación," *Diario Oficial*, Primera Sección, 76, no. 9 (May 13, 1941). In this article companies benefiting from said provisions must accept students who completed their studies at national universities or technical schools to do internships in their workshops or factories. Article 29 established the Office of Qualities and Industrial Standards.

54. Federico Barona, "Laboratorios Nacionales de Fomento Industrial," *Irrigación en México* 35, no. 4 (October–December 1945): 20–33.

55. See "Índice del volumen físico de la producción de la industria de transformación (1899–1959)," in Gonzalo Robles, "El Desarrollo Industrial," chap. 7 of *Ensayos sobre el desarrollo de México* (Mexico City: Fondo de Cultura Económica, 1982), 335.

56. Enrique Cárdenas, *La hacienda pública y la política económica 1929–1958* (Mexico City: COLMEX-Fondo de Cultura Económica, 1994), 102.

57. "Presentación," *Boletín del Instituto de Química* 1, no. 1 (December 1945): 3.

58. Fernando Orozco D. and Antonio Medinaveitia, "Estudio del Yacimiento de Salmueras Alcalinas del Valle de México," *Boletín del Instituto de Química* 1, no. 1 (1945): 6–25.

59. For a study of the steroid industry in Mexico, see J. Uriel Aréchiga Viramontes and Tomás Viveros, "Ingeniería Química. Breve historia," in *Cosmos*, 1:283–84.

60. *La política siderúrgica de México* (Mexico City: Ediciones del Doctorado en Administración Pública del IPN, 1976), 56.

61. Decree of President Miguel Alemán, October 21, 1948, qtd. in Antonio Acevedo Escobedo, *El Azufre en México* (Mexico City: Editorial Cultura, 1956), 41.

62. "Acuerdo de la Secretaría de Agricultura y Fomento del 27 de abril de 1942," *Diario Oficial* 121, no. 49, April 27, 1942,. See also Adolfo Orive Alba, *La irrigación en México* (México: Grijalbo, 1970).

63. Adolfo Orive Alba, "Política de Irrigación," in *Irrigación en México* 26, no. 1 (January–March 1945): 39–40.

64. Jesús Gracia, "La industria química de los fertilizantes," in Jesús Gracia et al., *Estado y fertilizantes (1760–1985)* (Mexico City: Fondo de Cultura Económica, 1988), 183.

CHAPTER 6: TECHNOCRATIC DIPLOMACY

1. Leo S. Rowe, "Closing Remarks of the President of the Academy," in *The Purposes and Ideals of the Mexican Revolution*, ed. Luis Cabrera et al. (Philadelphia: The American Academy of Political and Social Science, 1917), 30–31.

2. For more on states and making societies "legible," see Michel Foucault, *Discipline & Punish: The Birth of the Prison* (New York: Vintage Books, 1995); James C. Scott, *Seeing Like a State: How Certain Schemes to Improve the Human Condition Have Failed* (New Haven, CT: Yale University Press, 1999); Raymond B. Craib, *Cartographic Mexico: A History of State Fixations and Fugitive Landscapes* (Durham, NC: Duke University Press, 2004).

3. See Terry Shinn, Jack Spaapen, and Venni Krishna, eds., *Science and Technology in a Developing World* (New York: Springer, 1997); David Pretel and Lino Camprubí, *Technology and Globalisation: Networks of Experts in World History* (London: Palgrave Macmillan, 2018).

4. Andra B. Chasteen and Timothy W. Lorek, eds., *Itineraries of Expertise: Science, Technology, and Environment in Latin America's Long Cold War* (Pittsburgh: University of Pittsburgh Press, 2020), 3. Also see María Cecilia Zuleta, "Engineers' Diplomacy: The South American Petroleum Institute, 1941–1950s," in *Technology and Globalisation: Networks of Experts in World History*, ed. David Pretel and Lino Camprubí (London: Palgrave Macmillan, 2018), 341–70.

5. Ronald E. Doel, "Scientists as Policymakers, Advisors, and Intelligence Agents: Linking Contemporary Diplomatic History with the History of Contemporary Science," in *The Historiography of Contemporary Science and Technology*, ed. Thomas Söderqvist (New York: Hardwood Academic Publishers, 1997), 215–44; John Krige, *American Hegemony* (Cambridge, MA: MIT Press, 2006).

6. For a brief discussion on this topic, see Elmer Plischke, "The New Diplomacy," in *Modern Diplomacy: The Art and the Artisans*, ed. Elmer Plischke (Washington, DC: American Enterprise Institute for Public Policy Research, 1979), 57; Jean-Robert Leguey-Feilleux, *The Dynamics of Diplomacy* (Boulder, CO: Lynn Rienner, 2009), chap. 4.

7. Several historians of Mexico have examined Mexican diplomacy in the United States during the revolution. They have focused predominately on the rise and fall of Pancho Villa's popularity and the extensive diplomatic efforts of Venustiano Carranza to persuade US progressives to support his cause, to maintain peace with the United States, and to defend his nationalist positions. See Friedrich Katz, *The Secret War in Mexico: Europe, the United States, and the Mexican Revolution* (Chicago: University of Chicago Press, 1981); Douglas Richmond, *Venustiano Carranza's Nationalist Struggle, 1893–1920* (Lincoln: University of Nebraska Press, 1983), 189–218; Michael M. Smith, "Carrancista Propaganda and the Print Media in the United States: An Overview of Institutions," *The Americas* 52, no. 2 (October 1995): 155–74; Friedrich Katz, *The Life and Times of Pancho Villa* (Stanford, CA: Stanford University Press, 1998); Mark Cronlund Anderson, *Pancho Villa's Revolution by Headlines* (Norman: University of Oklahoma Press, 2000); Gabriel Rosenzweig, "Los diplomáticos mexicanos durante la Revolución: Entre el desempleo y exilio," *Historia Mexicana* 61, no. 4 (April–June 2012): 1461–523.

8. Helen Delpar, *Looking South: The Evolution of Latin Americanist Scholar-*

ship in the United States, 1850–1975 (Tuscaloosa: University of Alabama Press, 2008); Ricardo D. Salvatore, *Disciplinary Conquest: U.S. Scholars in South America, 1900–1945* (Durham, NC: Duke University Press, 2016), 2. Also see Ricardo D. Salvatore, "The Enterprise of Knowledge: Representational Machines of Informal Empire," in *Close Encounters of Empire: Writing the Cultural History of U.S.-Latin American Relations*, ed. Gilbert M. Joseph, Catherine L. Legrand, and Ricardo D. Salvatore (Durham, NC: Duke University Press, 1998), 69–104.

9. Salvatore, *Disciplinary Conquest*, 2.

10. Chasteen and Lorek, *Itineraries of Expertise*, 10–11.

11. Salvatore, *Disciplinary Conquest*, 10.

12. I am using "diplomatic" and "consular" loosely in this essay. Since during much of the revolution the US government did not officially recognize a Mexican government, revolutionaries serving in diplomatic roles had a myriad of titles, most commonly "commercial agent." For a discussion about studying diplomacy broadly, see Jeremy Black, *A History of Diplomacy* (London: Reaktion Books, 2010).

13. On the education of the Urquidi brothers, see Mílada Bazant, "Estudiantes mexicanos en el extranjero: El caso de los hermanos Urquidi," *Historia Mexicana* 36, no. 4 (April–June 1987): 739–58. For a broader overview of the rise of civil engineering education, see Lucía G. Santa Ana Lozada, "Orígines de la enseñanza de la ingeniería civil en México," in *Ingenioers de profesión, arquitectis de vocación: Venticinco protagonistas en la Arquitectura Mexicana del siglo xx*, ed. Ivan San Martín Córdova (Mexico City: UNAM, 2020), 35–49.

14. Charles Hale, *The Transformation of Liberalism in Late Nineteenth-Century Mexico* (Princeton, NJ: Princeton University Press, 2014), 5–6.

15. Hale, *Transformation of Liberalism*, 11–12.

16. J. Justin Castro, *Apostle of Progress: Modesto C. Rolland, Global Progressivism, and the Engineering of Revolutionary Mexico* (Lincoln: University of Nebraska Press, 2019), 17–20.

17. Jean Meynaud, *Technocracy* (New York: Free Press, 1968); Theodore Roszak, *The Making of a Counter Culture: Reflections of the Technocratic Society and Its Youthful Opposition* (Berkeley: University of California Press, 1968). That idea that technocracy is inherently antidemocratic has been challenged in more recent years, though the democracies that these scholars argue technocrats contributed to have characteristically been elite-run, limited democracies. Many of these works focus on economists of the later twentieth century. See Miguel Ángel Centeno, *Democracy within Reason: Technocratic Revolution in Mexico*, 2nd ed. (University Park: Pennsylvania State University Press, 1997); Eri Berstou and Daniele Caramani, eds., *The Technocratic Challenge to Democracy* (New York: Routledge, 2020).

18. Modesto C. Rolland, "Investigation Work into the Municipal City Governments and the Rural School System, Factories and Industrial Centres in the United States," in *Carranza and Mexico*, ed. Carlo de Fornaro (New York: Mitchell Kennerly, 1915), 107. I look at the influence of progressivism on this generation

of engineers, particularly Modesto Rolland, in my book *Apostle of Progress*. Carlo de Fornaro was born Carlo di Fornaro. For *Carranza and Mexico*, and some of his other publications, he changed the Italian "di" to the Spanish "de."

19. Although this was the perception promoted by many Carrancista propagandists, it was never wholly true. Zapata's forces, for example, possessed their own "Clandestine Wireless Office" with radio receivers that spied on Carrancista radio transmissions. See J. Justin Castro, *Radio in Revolution: Wireless Communications and State Power in Mexico, 1897–1938* (Lincoln: University of Nebraska, 2016), 68.

20. See Bazant, "Estudiantes mexicanos en el extranjero."

21. Bazant, "Estudiantes mexicanos en el extranjero," 744–46.

22. "United States Forced into Pledge to Rebels to Placate Carranza," *Washington Post*, July 1, 1914.

23. Peter Mathias and Sidney Pollard, *The Cambridge Economic History of Europe*, vol. 7, part 2, 2nd ed. (Cambridge: Cambridge University Press, 1987), 313; Bazant, "Estudiantes mexicano en el extranjero," 742, 750–51; Mílada Bazant, "La enseñanza y la práctica de la ingeniería durante el porfiriato," *Historia Mexicana* 33, no. 3 (January–March 1984): 277.

24. Lawrence Douglas Taylor Hansen, "Los Orígenes de la Fuerza Aérea Mexicana, 1913–1915," *Historia Mexicana* 56, no. 1 (July–September 2006): 189.

25. Francisco Urquidi, "'Let the United States Keep Hands Off Mexico and We'll Crush Huerta Soon,' Carranza's Agent Declares," *St. Louis Post-Dispatch*, April 19, 1914, 1.

26. "Who Don Francisco Is," *St. Louis Post-Dispatch*, April 19, 1914, 1.

27. Circular, September 9, 1914, exp. 17-20-33, Archivo Histórico Genearo Estrada de la Secretaría de Relaciones Exteriories, Mexico City (hereafter ASRE); [Name illegible] to Francisco Urquidi, September 18, 1914, exp. 17-20-33, ASRE.

28. "Carranza Defying Mexican Convention," *Atlanta Constitution*, October 25, 1914, 7; "Defied by Carranza," *Washington Post*, October 25, 1914, 2.

29. "Carranzistas Deny Forging Letters," *New York Times*, May 5, 1915, 5; "Wilson Demands Peace in Mexico," *New York Times*, June 3, 1915, 4; "Picks Man for Wilson," *New York Times*, June 5, 1915, 4.

30. Francisco Urquidi, qtd. in "Move to Recognize Carranza Weighed," *New York Times*, September 26, 1915, 10.

31. "Move to Recognize Carranza Weighed," *New York Times*, September 26, 1915, 10.

32. "Villa to Discharge His Consuls in U.S.," *Arkansas Gazette*, May 2, 1916, 4; Federico González Garza to Pedro Lamica, November 12, 1918, fondo CMXV, carp. 47, exp. 5679, f. 1, Centro de Estudios de Historia de México, Mexico City (hereafter CEHM).

33. "Mexican Board Filled," *Washington Post*, August 23, 1916, 1.

34. "Wilson Accepts Nomination on Party Record," *St. Louis Post-Dispatch*, September 3, 1916, A1.

35. Mary Austin, "What the Mexican Conference Really Means," *New York Times*, October 29, 1916, SM7.

36. "The Mexican Commission," *Outlook*, August 30, 1916, 1034.

37. "Mexican Commission," 1034.

38. Austin, "What the Mexican Conference Really Means."

39. Clarence C. Clendenen argued this point in 1961; see Clendenen, *The United States and Pancho Villa: A Study in Unconventional Diplomacy* (Ithaca, NY: Cornell University Press, 1961), 199.

40. Alberto Pani, *Apuntes autobiográficos* (Mexico City: Editorial Stylo, 1945), 127–50.

41. Pani published an English translation of this work in the United States; Pani, *Hygiene in Mexico: A Study of Sanitary and Educational Problems*, trans. Ernest L. De Gorgoza (New York: G. P. Putnam's Sons, 1917).

42. Pani, *Hygiene in Mexico*, 5–7.

43. Pani, *Hygiene in Mexico*, 120.

44. Pani, *Hygiene in Mexico*, 59, 69, 113.

45. Luis Cabrera, "The Mexican Revolution—Its Causes, Purposes and Results," in Cabrera et al., *Purposes and Ideals of the Mexican Revolution*, 1.

46. Douglas Richmond, *Venustiano Carranza's Nationalist Struggle, 1893–1920* (Lincoln: University of Nebraska Press, 1984), 191.

47. Ygnacio [Ignacio] Bonillas, "The Character and the Progress of the Revolution," in Cabrera et al., *Purposes and Ideals of the Mexican Revolution*, 20–21.

48. "Mexico Being Reconstructed, Says Peace Envoy," *New York Times*, November 5, 1916, SM9.

49. [George Weeks],"Telegraph, Post Office and Harbor Improvement," *Mexican Review*, November 1916, 3. For more on Weeks, see Michael M. Smith, "Gringo Propagandist: George F. Weeks and the Mexican Revolution," *Journalism History* 29, no. 1 (Spring 2003): 2–11.

50. "The Joint Mexican-American Commission," *Mexican Review*, October 1916, 3.

51. "Railway Traffic in Mexico," *Mexican Review*, November 1916, 6.

52. "The Man of the Hour in Yucatan," *Mexican Review*, October 1916, 13.

53. "Street Car Traffic in Mexico City," *Mexican Review*, November 1916, 13; "Wanted—An Oil Market," *Mexican Review*, October 1916, 11; "Promoting Industries in Guanajuato," *Mexican Review*, November 1916, 14; "Present Condition in Mexico," *Mexican Review*, December 1916, 4.

54. For an intelligent reflection on perceptions about technocratic governance across nations, see Magli Sarfatti Larson, "Notes on Technocracy: Some Problems of Theory, Ideology, and Power," *Berkeley Journal of Sociology* 17 (1972–1973): 1–34.

55. Alberto J. Pani, "The Sanitary and Educational Problems of Mexico," in Cabrera et al., *Purposes and Ideals of the Mexican Revolution*, 25.

56. Timothy G. Turner, *Bullets, Bottles, and Gardenias* (Dallas: South-West Press, 1935), 148.

57. William A. Maxwell, *A Quarter Century of Public School Development* (New York: American Book Company, 1912); William A. Maxwell, "Attitude of Parents toward Education," *Journal of Education* 77, no. 14 (April 3, 1913): 372–73; Wallace Sayre and Herbert Kaufman, *Governing New York City: Politics in the Metropolis* (Philadelphia: Russell Sage Foundation, 1960), 279, 311.

58. Modesto C. Rolland, "Investigation Work into the Municipal City Governments and the Rural School System, Factories and Industrial Centres in the United States," in de Fornaro, *Carranza and Mexico*, 110.

59. Clyde H. Tavenner, "Capital Comment," *Rock Island Argus*, August 17, 1914, 4; "Wilson's Stand Correct," *Bemidji Daily Pioneer*, August 17, 1914, 1; "The Reason Why," *The Menace*, August 22, 1914, 3; "Compulsory Education, Says Carranza, Is the Cure for Mexico's Ills," *The Sun*, September 20, 1914, sec. 3, 8.

60. "War with Mexico Averted," *Appendix to the Congressional Record*, 63rd Congress, Session 2 (Washington, DC: Government Printing Office, 1914), 1133; "Many Reforms Needed," *Arizona Sentinel*, August 20, 1914, 4.

61. "The Progress of the World," *American Review of Reviews* 40, no. 6 (December 1909): 661; "Mexico Is Portrayed in Books," *Publisher's Weekly*, July 8, 1916, 102. For Rolland's views on municipal governance, see Modesto C. Rolland, "Investigation Work into the Municipal City Governments and the Rural School System, Factories and Industrial Centres in the United States," in de Fornaro, *Carranza and Mexico*, 106–13; Modesto C. Rolland, *El desastre municipal en la República Mexicana*, 2nd ed. (Mexico City: I. Molina M., 1939).

62. Joel Menendez Ruiz, "Modesto C. Rolland y el primer libro sobre el municipio libre mexicano," in *Municipalistas y municipalismo en México*, ed. Carlos Reta Martínez (Mexico City: Instituto Nacional de Administración Pública, 2018), 261–67.

63. A large selection of the actual articles can be viewed in the Hathi Trust Digital Library, www.hathitrust.org.

64. M. C. Rolland to Venustiano Carranza, January 16, 1917, fondo XXI, carp. 109, exp. 12479, f. 2, CEHM.

65. "Many Problems Solved in Yucatan," *New York Times*, October 1, 1916, SM3.

66. Paul U. Kellogg, "A New Era of Friendship for North America," *The Survey*, July 15, 1916, 415–17.

67. M. C. Rolland, "A Trial of Socialism in Mexico: What the Mexicans Are Fighting For," *Forum* 56 (July–December 1916): 79–91.

68. Katz, *Life and Times of Pancho Villa*, 319–20;324–26 (for Villa's work with Hollywood).

69. See Anderson, *Pancho Villa's Revolution by Headlines*.

70. Carranza too relied significantly on foreign agents, including lawyer

Sherburne G. Hopkins, attorney Charles A. Douglas, and newspaperman George Weeks. But Carranza ultimately put more trust, stock, and numbers into officials of his government. See Smith, "Carrancista Propaganda and the Print Media in the United States."

71. Katz, *Life and Times of Pancho Villa*, 317–18.

72. Federico González Garza, *La Revolución Mexicana: Mi contribución político-litería* (Mexico City: A. Del Bosque, 1936), 369–72.

73. Clendenen, *United States and Pancho Villa*, 158, 201–2.

74. Lincoln Steffens, qtd. in P. Edward Haley, *Revolution and Intervention: The Diplomacy of Taft and Wilson with Mexico* (Cambridge, MA: MIT Press, 1970), 177.

75. "Wilson Sends Demand to Mexican Leaders to Reach Agreement Now," *Washington Post*, June 3, 1915, 1; "Villa to Protect All," *Washington Post*, September 21, 1915, 3. Villa and Zapata were not opposed to education; they both had a sincere appreciation for increased education in Mexico, but Carranza's agents in the United States made a greater point of it and actively sought to influence and incorporate US educators.

76. Robert Lansing, qtd. in "Carranza Recognized by Pan-America," *Atlanta Constitution*, October 10, 1915, 1.

CHAPTER 7: PUNITIVE ENGINEERING AND MILITARY MODERNIZATION

I would like to thank Paul Gillingham, Terry Rugeley, Justin Castro, and the archivists at the Fideicomiso Archivos de Plutarco Elías Calles y Fernando Torreblanca for their time and support on this research project. I appreciate the organizers of the Patton and Pancho Conference and the International Conference of Mexican Historians for providing me space to share and refine the ideas above.

Epigraph: John F. Madden, "Thanks to Villa," *Quartermaster Review*, October 1921, 31.

1. "A Report on a Two Hundred Mile Experimental Cavalry March," *Journal of the United States Cavalry Association* 27 (June 1917): 27; Charles H. Harris III and Louis R. Sadler, *The Great Call-Up: The Guard, the Border, and the Mexican Revolution* (Norman: University of Oklahoma Press, 2016), 81.

2. Floyd P. Gibbons, "12,000 Await Trumpet for Villa Pursuit—Pershing Moves Command to Columbus; Villa Takes to the Hills," *Chicago Daily Tribune*, March 14, 1916, 1; Frank Elser, "Bandit Not Surrounded: All Going Well, Pershing Tells Times, but End Not in Sight," *New York Times*, March 16, 1916, 1; "Gen. Wood to Send 130,000 Soldiers to Mexican Line . . . 24,000 Already on the Way . . . Only 58,000 Now Left in the Mobilization Camps Under Wood's Jurisdiction," *New York Times*, July 3, 1916, 1.

3. For major works, see Wendy Waters, "Re-mapping the Nation: Road Building as State Formation in Post-revolutionary Mexico, 1925–1940" (PhD diss.,

University of Arizona, 1999); Waters, "Remapping Identities: Road Construction and Nation Building in Postrevolutionary Mexico," in *The Eagle and the Virgin: Nation and Cultural Revolution in Mexico, 1920–1940*, ed. Mary Kay Vaughn and Stephen E. Lewis (Durham, NC: Duke University, 2006), 221–42; Michael Bess, "Revolutionary Paths: Motor Roads, Economic Development, and National Sovereignty in 1920s and 1930s Mexico," *Mexican Studies/Estudios Mexicanos* 32, no. 1 (Winter 2016): 56–82; Michael K. Bess, *Routes of Compromise: Building Roads and Shaping the Nation in Mexico, 1917–1952* (Lincoln: University of Nebraska Press, 2017); Héctor Mendoza Vargas, "El automóvil y los mapas en la integración del territorio Mexicano, 1929–1962," *Investigaciones Geográficas, Boletín del Instituto de Geografía* (UNAM) 88 (2015): 91–108; Luis Anaya Merchant, "Guerra, automóviles y carretera: La influencia norteamericana y el mercado automotriz mexicano en la 'reconstrucción' posrevolucionaria," *Boletín* (FAPECFT) 73, no. 1 (May–August 2013): 1–32; Víctor Manuel Gruel Sández, "La Comisión Nacional de Caminos, 1925–1932," *Boletín* (FAPECFT) 80, no. 2 (September–December 2015): 1–47.

4. On the notion of contact zones, or places or moments wherein two or more different sets of actors come into contact, see Mary Louise Pratt, "Arts of the Contact Zone," *Profession* (1991): 33–40, https://www.jstor.org/stable/25595469.

5. Prieto, *The Mexican Expedition, 1916–1917* (Washington, DC: Center of Military History, United States Army, 2016), 9. Cultural historians have noted the persistence of informal forms of cooperation. Helen Delpar traces how US political leftists, artists, and academics traveled to Mexico in search of cultural authenticity and in many cases forged significant relationships with Mexican elites; Delpar, *The Enormous Vogue of Things Mexican: Cultural Relations between the United States and Mexico, 1920–1935* (Tuscaloosa: University of Alabama Press, 1992). Megan Threlkeld highlights how the members of four major US women's organizations, including the Women's Peace Society and Women's International League for Peace and Freedom, "worked to establish contacts, to exchange information with Mexican women about their restive programs"; Threlkeld, *Pan American Women: U.S. Internationalists and Revolutionary Mexico* (Philadelphia: University of Pennsylvania Press, 2014), 19.

6. Prieto, *Mexican Expedition*; Harris and Sadler, *Great Call-Up*.

7. Susan Gauss, *Made in Mexico: Regions, Nation, and the State in the Rise of Mexican Industrialism, 1920s–1940s* (University Park: Pennsylvania State University Press, 2010), 27.

8. Thomas Rath, *The Myth of Demilitarization in Postrevolutionary Mexico, 1920–1960* (Chapel Hill: University of North Carolina Press, 2013), 15–16.

9. Stephen Skowronek, *Building a New American State: The Expansion of National Administrative Capacities, 1877–1920* (Cambridge: Cambridge University Press, 1982), 8.

10. Stephen Neufeld, *The Blood Contingent: The Military and the Making of*

Modern Mexico, 1876–1911 (Albuquerque: University of New Mexico Press, 2017), 131–32, 232, 238.

11. "Revista mensual de Bellas Artes e Ingeniería," *El Arte y la Ciencia*, Archivo Histórico de la Secretaría de la Defensa Nacional, Mexico City (hereafter as AHSDN).

12. A Drexel University and Colegio Militar graduate, General Mondragón sold his handmade weapons overseas. Neufeld, *Blood Contingent*, 239.

13. Tactless yet symbolic, Mexican officials exhibited the head of Apache leader Juan Antonio at the 1889 World's Fair to embody national order and progress. Neufeld, *Blood Contingent*, 246; Tenorio-Trillo, *Mexico at the World's Fairs: Crafting a Modern Nation* (Berkley: University of California Press, 1996).

14. Lawrence J. Fleming, "The Automatic Pistol in the Punitive Expedition," *Journal of the United States Cavalry Association* 27 (April 1917): 497–514.

15. Villa's sense of betrayal came from the fact that he considered himself the protector of US property in revolutionary Chihuahua. In addition to various businessmen, Villa also had personal relations with US military personnel. Alan Knight, *U.S.-Mexican Relations, 1910–1940: An Interpretation* (San Diego: Center for U.S.-Mexican Studies, 1987), 22; Friedrich Katz, *The Life and Times of Pancho Villa* (Stanford, CA: Stanford University Press, 1998).

16. Harris and Sadler, *Great Call-Up*; Joseph Brinker, "Civilian Motor Trucks as Army Supply Trains," *Scientific American*, April 22, 1916, 423, 437–38.

17. Harris and Sadler, *Great Call-Up*, 82–83, 159.

18. The army commissioned 716 Packard vehicles alone. Both companies discouraged drivers from participating in the expedition directly; Harris and Sadler, *Great Call-Up*, 303.

19. Prieto, *Mexican Expedition*, 49; Harris and Sadler, *Great Call-Up*, 60.

20. Josefina Zoraida Vázquez and Lorenzo Meyer, eds., *México frente a Estados Unidos (Un ensayo histórico, 1776–1993)* (Mexico City: Fondo de Cultura Económica, 1995), 138.

21. "Place Curb on Funston; Must Not Occupy Towns—Washington Official Fail to Give Him 'Free Hand' Promised at Start of Expedition," *Chicago Daily Tribune*, 2; "Gave Army New Life," *Washington Post*, August 13, 1916, 10.

22. "Mexicans Astounded by Rapid Advance of American Punitive Expedition," *Los Angeles Times*, March 24, 1916.

23. Clarence Lininger, "The Type of Cavalry Horse for Campaign," *Journal of the United States Cavalry Association* 27 (April 1917): 582.

24. Mitchell Yockelson, *Forty-Seven Days: How Pershing's Warriors Came of Age to Defeat the German Army in World War 1* (New York: Caliber/Random House, 2016).

25. Alan Knight, *The Mexican Revolution*: Counter-revolution and Reconstruction (Lincoln: University of Nebraska Press, 1986), 2:348; Prieto, *Mexican Expedition*, 25. David Nugent estimates that up to 20 percent of the men who

attacked Columbus with Pancho Villa were from Namiquipa; Nugent, *Spent Cartridges of Revolution: An Anthropological History of Namiquipa, Chihuahua* (Chicago: University of Chicago Press, 1993), 82.

26. Terry Keith Derouchey, "The U.S. Punitive Expedition in Chihuahua 1916–1917: The Legacy of U.S. Military Intervention" (MA thesis, University of Texas at Austin, 1989), Rare Books, 55, Nettie Lee Benson Library, University of Texas, Austin (hereafter NLBL).

27. Madden, "Thanks to Villa," 29; James A. O'Connor, "Road Work in Mexico with the Punitive Expedition," *Professional Memoir (Military Engineer)*, December 1917, 338; "Military Road in Desert: Made from Villa's Trail by Advancing American Troops," *New York Times*, March 23, 1916.

28. Prieto, *Mexican Expedition*, 49.

29. According to Captain Ernest Graves of the Corps of Engineers, it was only after a couple of terribly close calls delivering food that funding was provided for the necessary road-building materials, Graves, "Road Work on the Punitive Expedition into Mexico," *Military Engineer*, November–December 1917, 676; O'Connor, "Road Work in Mexico with the Punitive Expedition," 330.

30. Prieto, *Mexican Expedition*, 50.

31. Ernest Graves, "Military Road Building Not a Science—Only a Job," *Military Engineer*, March 1919, 161.

32. O'Connor, "Road Work in Mexico with the Punitive Expedition," 330.

33. However, time was of the essence, timber was scarce, and the dry weather problems were as bad as any rainy season complications ahead, leaving adobe roads as the optimal option; Graves, "Road Work on the Punitive Expedition into Mexico," 680.

34. Graves, "Road Work on the Punitive Expedition into Mexico," 681.

35. O'Connor, "Road Work in Mexico on the Punitive Expedition," 330.

36. Harris and Sadler, *Great Call-Up*, 159; Prieto, *Mexican Expedition*, 49.

37. H. D. Church, "Military Motor Truck Design," *Horseless Age*, May 1917, 2–6; Fleming, "Automatic Pistol in the Punitive Expedition," 498.

38. Harris and Sadler, *Great Call-Up*, 30.

39. Prieto, *Mexican Expedition*, 24.

40. As was the case in other foreign wars, manifest destiny, masculinity, and racism fueled the popular draw of the militia.

41. Harris and Sadler, *Great Call-Up*, 171–72, 339–340, 353–54.

42. Nugent, *Spent Cartridges of Revolution*, 82.

43. Soldiers and civilians were the laborers. By "local," Graves meant the need for one engineer per detachment. The chairman of the Committee of Emergency Construction, Colonel W. A. Starrett called this "the theory of decentralization," in which engineers indicate standards in their main offices with the expectation to later "adapt them to local conditions," Colonel W. A. Starrett, "The Construction Division of the United States Army: The Vast Machine that Attends to Emergency

Building for the War Department," *Scientific American*, September 28, 1918, 252; O'Connor, "Road Work in Mexico with the Punitive Expedition," 343.

44. Graves, "Military Road Building Not a Science," 162; and Graves, "Road Work on the Punitive Expedition into Mexico," 166.

45. "The rain question was an interesting one. Without water, the roads soon became impassable and with too much of it was almost universally feared that the traffic would cease." Too little water was an actual problem, but having too much was only a perceived one, albeit universally so. O'Connor, "Road Work in Mexico with the Punitive Expedition," 330.

46. Prieto, *Mexican Expedition*, 68.

47. Daniel T. Rodgers, *Atlantic Crossings: Social Politics in a Progressive Age* (Cambridge, MA: Harvard University Press, 1998).

48. Frank B. Elser, "War on Bad Roads Keeps Pershing Busy: Approach of Rainy Season Pushes Work of Building—Bandit Leader Shot," *New York Times*, June 11, 1916; "Army Road Building with Jeffery 'Quads,'" *Good Roads*, October 7, 1916, 166; Captain Henry J. Reilly, "Mexican Crisis a Real Lesson," *Los Angeles Times*, December 8, 1916.

49. O'Connor, "Road Work in Mexico with the Punitive Expedition," 326.

50. J. C. Burton, "Good Roads Vital Factor in National Defense Campaign," *Chicago Daily Tribune*, June 4, 1916.

51. The Liberty Trucks offer a good example. Made by the Transport Section of the Quartermaster Corps and the Society of Automotive Engineers, 7,500 went directly to France. Prieto, *Mexican Expedition*, 68.

52. "Motor Truck's Power Proved: Speaker Tells How Pershing Supply Service Worked," *Los Angeles Times*, March 18, 1917.

53. Raymond H. Thompson, "How Pancho Villa and Emil Haury Established Highway Salvage Archaeology in Arizona," *Journal of the Southwest* 46 (Spring 2004): 121–22.

54. Earl Swift, *The Big Roads: The Untold Story of the Engineers, Visionaries, and Trailblazers Who Created the American Superhighways* (New York: Mariner Books, 2012), 76.

55. Tammy Ingram, *Dixie Highway: Road Building and the Making of the Modern South, 1900–1930* (Chapel Hill: University of North Carolina Press, 2014), 113.

56. C. H. Claudy, "Behind the Cantonments: Picking the Men for America's Greatest Emergency Engineering Feat," *Scientific American*, September 1, 1917, 156.

57. Colonel W. A. Starrett, "The Construction Division of the United States Army," *Scientific American*, September 28, 1918, 252.

58. "The National Army Cantonments: Building Sixteen Towns of 40,000 Inhabitants in Ninety Days," *Scientific American*, December 1, 1917, 406.

59. Colonel W. A. Starrett, "The Government's Vast Building Program: Two

Billion Dollars Worth of Construction Required for Our Army," *Scientific American*, September 7, 1918, 186–87, 197–98.

60. Claudy, "Behind the Cantonments."

61. "Some Motor Truck Economics: National Standardized Roads for the Nation's Motor Traffic," *Scientific American*, December 28, 1918, 530.

62. Bruce E. Seely, *Building the American Highway System: Engineers as Policy Makers* (Philadelphia: Temple University Press, 1988), 58.

63. Archivo Joaquín Amaro (hereafter AJA) is located in the Fideicomiso Archivos Plutarco Elías Calles y Fernando Torreblanca (hereafter FAPECFT), along with the Fondo Obregón (hereafter FO), Fondo Plutarco Elías Calles (hereafter FPEC), and Archivo Fernando Torreblanca (hereafter AFT), all in Mexico City. The Colección Documental Embajada de los Estados Unidos en México (hereafter CDEEUM) is a separate section of AJA.

64. Luis Montes de Oca, "Mexico Is America's Market," *Mexico: Commercial Interest of the United States Being Sacrificed by Diplomatic Deadlock*, 1922, 2–3, Hoover Presidential Archive, West Branch, Iowa.

65. American army officer Captain Laval traveled to Mexico City to offer the War Department deals on airplanes, artillery, cavalry equipment, and ammunition trucks. R. M. Campbell, "Weekly Report—Military," May 12, 1919, Agregado Militar de Estados Unidos: Reportes [Informes], Inv. 10 Serie 040101 Leg. 5/10 (1919), ff. 200, CDEEUM. In 1918 Ford reached out to Carranza about the development of a tractor factory in Mexico City; Campbell, on an *Excélsior* article (no title given), July 29, 1918, Agregado Militar de Los Estados Unidos: Reportes, 030101/Leg. 2/7/ Inv.3, CDEEUM. The Ford Model-T was built for the US South but gained traction in Latin America. Ford tractors were especially popular in Mexico. Michael L. Berger, *The Devil's Wagon in God's Country: The Automobile and Social Change in Rural America, 1893–1929* (Hamden, CT: Archon Books, 1979), 16, 40–41, 51.

66. Justin Castro, *Radio in Revolution: Wireless Technology and State Power in Mexico, 1897–1938* (Lincoln: University of Nebraska Press, 2016), 71, 129, 131–32.

67. What Edwin Lieuwen calls the Jacobin phase of the revolution was an early military-technocratic stage. Revolutionary Mexico City under Obregón vaguely resembled Robespierre's "Manufacture Paris of 1794" in technocracy, not terror; Lieuwen, *Mexican Militarism: The Political Rise and Fall of the Revolutionary Army, 1910–1940* (Albuquerque: New Mexico University Press, 1968), 43. Ken Alder shows that at the height of the French Revolution in 1794, revolutionaries transformed Paris into a gun-making factory of exceptional proportions; Alder, *Engineering the Revolution: Arms and Enlightenment in France, 1763–1815* (Chicago: University of Chicago Press, 1997), 253–91.

68. Lieuwen, *Mexican Militarism*, 64, 67, 72, 74.

69. For the military elite (generals, colonels, and other officers) the reserve meant two years of half-pay and the retention of rank and uniform; Lieuwen, *Mexican Militarism*, 67.

70. Lieuwen, *Mexican Militarism*, 153.

71. Jürgen Buchenau, *In the Shadow of the Giant: The Making of Mexico's Central America Policy, 1876–1930* (Tuscaloosa: University of Alabama Press, 1996), 142, 146. John J. Dwyer shows that Obregón redistributed only 4,142,355 acres of land from 1920 to 1924; Dwyer, *The Agrarian Dispute: The Expropriation of American-Owned Rural Land in Postrevolutionary Mexico* (Durham, NC: Duke University Press, 2008), 22.

72. George M. Russell, "Weekly Report—Works IX," February 16, 1921, División de Inteligencia Militar: Informes: Inv. 19 Exp. 060201 Leg. 2/6, ff. 83, CDEEUM.

73. George M. Russell, "Weekly Report—Works," February 16, 1921, División de Inteligencia Militar: Informes: Inv. 19 Exp. 060201 Leg. 1/6, ff. 18; Russell, "Weekly Report—Misc.," February 15, 1922, Agregado Militar de Estados Unidos: Informes: Inv. 23 Exp. 070101 Serie 1922 Leg. 1/11, ff. 26, both in CDEEUM.

74. George M. Russell, "Weekly Report—Military Agricultural Colonies," June 14, 1922, Agregado Militar de Estados Unidos: Informes: Inv. 23 Exp. 070101 Serie 1922 Leg. 5/11, ff. 160, CDEEUM.

75. George M. Russell, "Weekly Report—Military Colonies," August 5, 1921, División de Inteligencia Militar: Informes: Inv. 19 Exp. 060201 Leg. 4/6, ff. 148; Russell, "Weekly Report—Military," October 20, 1921, División de Inteligencia Militar: Informes: Inv. 19 Exp. 060201 Leg. 5/6, ff. 187, both in CDEEUM.

76. George M. Russell, "Weekly Report—Comment," August 10, 1921, División de Inteligencia Militar: Informes: Inv. 19 Exp. 060201 Leg. 4/6, ff. 152, CDEEUM.

77. As late as December the government released a strategy for building twenty-two military hospitals; George M. Russell, "Weekly Report–Military," December 21, 1921, División de Inteligencia Militar: Informes: Inv. 19 Exp. 060201 Leg. 5/6, ff. 215, CDEEUM.

78. George M. Russell, "Weekly Report—Freight Conditions," June 1, 1921, División de Inteligencia Militar: Informes: Inv. 19 Exp. 060201 Leg. 2/6, ff. 101; Russell, "Weekly Report—Economic," December 14, 1921, División de Inteligencia Militar: Informes: Inv. 19 Exp. 060201 Leg. 5/6, ff. 216, both in CDEEUM.

79. Bess, *Routes of Compromise.*

80. Faustino Roel, "Primer Congreso Nacional de Caminos Comisión Organizadora a Director de La Escuela de Minería," June 24, 1921, Caja 1921/1/358/No. 7, Archivo Histórico de la Escuela Nacional de Ingenieros, Mexico City (hereafter AH-ENI).

81. Faustino Roel to Mariano Moctezuma, June 22, 1921, Caja 1921/3/360/No. 1; Mariano Moctezuma to Faustino Roel, June 22, 1921, Caja 1921/3/360/No. 1, both in AH-ENI.

82. George M. Russell, "Weekly Report—Roads," June 27, 1921, División de

Inteligencia Militar: Informes: Inv. 19 Exp. 060201 Leg. 3/6, ff. 119, CDEEUM; "Las Carreteras Son una Grande Mejora Pública: Una Verdadera Riqueza Nacional son los Caminos que Recorrerán los Automóviles," *Excélsior*, July 31, 1921, 2; "La Sesión de Ayer en el Congreso de Caminos," *El Universal*, September 9, 1921; Russell, "Weekly Report—Wireless," August 17, 1921, División de Inteligencia Militar: Informes: Inv. 19 Exp. 060201 Leg. 4/6, ff. 158, CDEEUM; Russell, "Weekly Report—Military," October 20, 1921.

83. George M. Russell, "Weekly Report—Wireless," August 17, 1921; Russell, "Weekly Report—Military," October 20, 1921.

84. Russell, "Weekly Report—Misc.," August 12, 1921, División de Inteligencia Militar: Informes: Inv. 19 Exp. 060201 Leg. 4/6, ff. 154, CDEEUM.

85. George M. Russell, "Weekly Report—Misc.," September 9, 1921, División de Inteligencia Militar: Informes: Inv. 19 Exp. 060201 Leg. 4/6, ff. 169, CDEEUM.

86. George M. Russell, "Lack of Progressiveness. Resistance to new Highways," November 11, 1925, Agregado Militar de Estados Unidos: Informes: Inv. 40 Exp. 100203 Serie 1925 Leg. 5/7, CDEEUM; Jürgen Buchenau, *The Last Caudillo: Álvaro Obregón and the Mexican Revolution* (Malden, MA: Wiley-Blackwell, 2011), 112.

87. Sometimes it did not. Appointed as the minister of education, the philosopher and educator José Vasconcelos received a generous budget to stimulate national access to public education; Buchenau, *Last Caudillo*, 123.

88. George M. Russell, "Weekly Report—Misc. Military Information," February 1, 1922, Agregado Militar de Estados Unidos: Informes: Inv. 23 Exp. 070101 Serie 1922 Leg. 1/11, ff. 15, CDEEUM.

89. George M. Russell, "Weekly Report—Military," November 23, 1921, División de Inteligencia Militar: Informes: Inv. 19 Exp. 060201 Leg. 5/6, ff. 206, CDEEUM.

90. George M. Russell, "Transportation-Roads," March 15, 1922, Agregado Militar de Estados Unidos: Informes: Inv. 23 Exp. 070101 Serie 1922 Leg. 3/11, ff. 76; Russell, "Weekly Report—Savings Bank for Soldiers," September 30, 1922, Agregado Militar de Estados Unidos: Informes: Inv. 23 Exp. 070101 Serie 1922 Leg. 9/11, ff. 241; Russell, "Weekly Report—Organization," November 8, 1922, Agregado Militar de Estados Unidos: Informes: Inv. 23 Exp. 070101 Serie 1922 Leg. 10/11, ff. 280, all in CDEEUM.

91. George M. Russell, "Weekly Report—Equipment and Munitions," February 15, 1922, Agregado Militar de Estados Unidos: Informes: Inv. 23 Exp. 070101 Serie 1922 Leg. 1/11, ff. 26, CDEEUM.

92. George M. Russell, "Mexico: General: Routes and Highways," March 29, 1922, Agregado Militar de Estados Unidos: Informes: Inv. 23 Exp. 070101 Serie 1922 Leg. 2/11, ff. 68; Russell, "Weekly Report—Transportation," November 22, 1922, Agregado Militar de Estados Unidos: Informes: Inv. 23 Exp. 070101 Serie

1922 Leg. 10/11, ff. 303; Russell, "Weekly Report—Military Agricultural Colonies," December 20, 1922, Agregado Militar de Estados Unidos: Informes: Inv. 23 Exp. 070101 Serie 1922 Leg. 11/11, ff. 335, all in CDEEUM. My periodization parallels Michael Bess's description of delays in 1922; Bess, *Routes of Compromise*, 27.

93. In Chihuahua in particular, "capitalist interests" and the promise of an automobile road from Ciudad Chihuahua to Ciudad Juárez provided worked for three thousand troops; George M. Russell, "Weekly Report—Automobile Road," June 3, 1921, División de Inteligencia Militar: Informes: Inv. 19 Exp. 060201 Leg. 3/6, ff. 105, CDEEUM.

94. George M. Russell, "Weekly Report—Routes and Highways," February 22, 1922, Agregado Militar de Estados Unidos: Informes: Inv. 23 Exp. 070101 Serie 1922 Leg. 1/11, ff. 35; Russell, "Weekly Report—Routes and Highways," February 23, 1922, Agregado Militar de Estados Unidos: Informes: Inv. 23 Exp. 070101 Serie 1922 Leg. 1/11, ff. 23, both in CDEEUM.

95. Steven Haber, *Industry and Underdevelopment: The Industrialization of Mexico, 1890–1940* (Stanford, CA: Stanford University Press, 1989), 138–40, 156.

96. One newspaper even suggested that cavalry horses were dying in droves in Mexico's central valley because of "neglectful forage officers"; George M. Russell, "Weekly Report—Misc. (Military Instruction Camps)," August 2, 1922, Agregado Militar de Estados Unidos: Informes: Inv. 23 Exp. 070101 Serie 1922 Leg. 7/11, CDEEUM; Lieuwen, *Mexican Militarism*, 69.

97. George M. Russell, "Weekly Report—Who's Who," March 1922, Agregado Militar de Estados Unidos: Informes: Inv. 23 Exp. 070101 Serie 1922 Leg. 2/11, ff. 72, CDEEUM.

98. George M. Russell, "Weekly Report—Horse Buying," July 18, 1921, División de Inteligencia Militar: Informes: Inv. 19 Exp. 060201 Leg. 3/6, ff. 134; Russell, "Weekly Report—Remounts," June 17, 1921, División de Inteligencia Militar: Informes: Inv. 19 Exp. 060201 Leg. 3/6, ff. 112, both in CDEEUM.

99. Russell, "Weekly Report—Horse Buying," August 15, 1921, División de Inteligencia Militar: Informes: Inv. 19 Exp. 060201 Leg. 4/6, ff. 156, CDEEUM.

100. Knight, *U.S.-Mexican Relations, 1910–1940*, 96.

101. Notably, Arizona had only two honorary commissions (Miami, AZ and Gibson, AZ); George M. Russell, "Weekly Report—Mexican Honorary Commissions," August 17, 1921, División de Inteligencia Militar: Informes: Inv. 19 Exp. 060201 Leg. 4/6, ff. 158, CDEEUM.

102. Packard Automobiles, "Lista de accesorios para la reparación de un automóvil marca 'Packard Twin Six' presentada al Sr. F. Torreblanca," 1922, Packard, Automóvil (1922–1923), Fondo 03 Serie 02.04 Exp. P-8/151 Inv. 324 Leg. 1; Fernando Torreblanca to E. D. Ruíz, August 3, 1922, Packard, Automóvil (1922–1923), Fondo 03 Serie 02.04 Exp. P-8/151 Inv. 324 Leg. 1, ff. 11, both in Archivo Fernando Torreblanca–Fondo Fernando Torreblanca, Mexico City; Citizens Auto

Company, "Invoice: Citizens Auto Company (San Antonio, TX)," September 14, 1922, Fracturas y Documentos suplementarios: Inv. 63 Exp. 3 Serie 0105 Leg. 1/10, ff. 52, AJA.

103. General Angel Estrada, for instance, unofficially met with US military soldiers to purchase arms in 1922; George M. Russell, "Weekly Report–Sonora," January 25, 1922, Agregado Militar de Estados Unidos: Informes: Inv. 23 Exp. 070101 Serie 1922 Leg. 1/11, ff. 8, CDEEUM.

104. George M. Russell, "Weekly Report—Economic (Military Road for Use and Jobs)," January 18, 1922, Agregado Militar de Estados Unidos: Informes: Inv. 23 Exp. 070101 Serie 1922 Leg. 1/11, CDEEUM.

105. George M. Russell, "Weekly Report—Routes and Highways," June 28, 1922, Agregado Militar de Estados Unidos: Informes: Inv. 23 Exp. 070101 Serie 1922 Leg. 5/11, CDEEUM.

106. George M. Russell, "Weekly Report—Who's Who," January 25, 1922, Agregado Militar de Estados Unidos: Informes: Inv. 23 Exp. 070101 Serie 1922 Leg. 1/11, ff. 11; Russell, "Weekly Report—Lower California," February 15, 1922, Agregado Militar de Estados Unidos: Informes: Inv. 23 Exp. 070101 Serie 1922 Leg. 1/11, ff. 27, both in CDEEUM.

107. B. F. Fly to José Inocente Lugo, April 10, 1922, Lugo, José Inocente (Lic.) 1922, Inv. 3316 Exp. 127 Gav. 49 Leg. 2/12, FPEC.

108. José Inocente Lugo to Plutarco Elías Calles, June 15, 1922; Lugo to Calles, June 15, 1922; "Darlington sugiere una zona neutral para la carretera Calexico-Yuma que se proyecta," *Calexico Chronicle*, June 2, 1922, all in FPEC.

109. George M. Russell, "Weekly Report—Who's Who (Gen. Covarrubias & Ing. Duvallon)," June 14, 1922, Agregado Militar de Estados Unidos: Informes: Inv. 23 Exp. 070101 Serie 1922 Leg. 5/11, ff. 157; Russell to A.C. of S., G-2., "Cavalry Training Manuals," December 8, 1923, Agregado Militar de Estados Unidos: Correspondencia Inv. 26 Exp. 080101 Serie 1923 Leg. 3/3, ff. 188; Russell, "The Mexican Military College," December 7, 1923, Agregado Militar de Estados Unidos: Correspondencia Inv. 26 Exp. 080101 Serie 1923 Leg. 3/3. ff. 186–87; Russell, "Who's Who: General Victor Hernandez Covarrubias," May 10–17, 1922, Agregado Militar de Estados Unidos: Informes: Inv. 23 Exp. 070101 Serie 1922 Leg. 4/11 ff. 119–20, all in CDEEUM.

110. F. Le. J. Parker, "Invitation for Mexican Officers to attend U.S. Service Schools," *Excélsior*, May 8, 1923, Agregado Militar de Estados Unidos: Correspondencia Inv. 26 Exp. 080101 Serie 1923 Leg. 1/3, CDEEUM.

111. By October, it was clear that "the boom expected to materialize upon 'Recognition' [had] not arrived." Strapped for cash, local merchants began to refuse to "extend further credit to the various government supply agencies." "General Situation–Economic (Vickers-Armstrong Liquidation)," October 27, 1923, Agregado Militar de Estados Unidos: Informes: Inv. 27 Exp. 080102 Serie 1923 Leg. 7/7; "General Situation—Economic (Military Manufacturing Plants in Troubles),"

November 10, 1923, Agregado Militar de Estados Unidos: Informes: Inv. 27 Exp. 080102 Serie 1923 Leg. 7/7, ff. 413, both in CDEEUM.

112. Edward R. Stone, "Weekly Report—Military Agricultural Colonies (Success in Veracruz and Michoacán)," August 19, 1923, División de Inteligencia Militar: Informes: Inv. 29 Exp. 080201 Serie 1923 Leg. 4/6, ff. 210, CDEEUM.

113. George M. Russell, "Weekly Report—Misc.—Flying Columns," May 23, 1923, División de Inteligencia Militar: Informes: Inv. 29 Exp. 080201 Serie 1923 Leg. 3/6, ff. 126, CDEEUM.

114. Seen traveling between Nuevo Leon, Coahuila, and Hidalgo, General Mensses's "flying columns" of the 60th Cavalry were particularly noteworthy; George M. Russell, "Weekly Report—Movements," June 3, 1921, División de Inteligencia Militar: Informes: Inv. 19 Exp. 060201 Leg. 3/6 (1921), CDEEUM. Leo J. Daugherty III defines "flying columns," as an assault formation. He also argues that the occupation of Veracruz changed US urban occupation and army engineering.. Daugherty, *Counterinsurgency and the United State Marine Corps*, vol. 1, *The First Counterinsurgency Era, 1899–1945* (Jefferson, NC: McFarland, 2015), 47.

115. Edward R. Stone, "Weekly Report—Troops Building Roads," October 10, 1923, División de Inteligencia Militar: Informes: Inv. 29 Exp. 080201 Serie 1923 Leg. 5/6, CDEEUM.

116. Buchaneu, *Last Caudillo*, 131, 134.

117. This strategy eerily foreshadows how Joaquín Amaro managed violence during the Cristero Revolt.

118. Obregón to Gral. F. R. Serrano, March 17, 1924, Sánchez/A/III/1–47/15–2782, AHSDN.

119. Gral. Rafael Sánchez to Obregón, March 16, 1924, Sánchez/A/III/1–47/15–2782, ff. 607; Gral. Rafael Sánchez to Obregón, March 21, 1924, Sánchez/A/III/1–47/15–2782, ff. 651, both in AHSDN.

120. "Nuevo Jefe de los Campos de Concentración del Ejército," *El Universal*, December 9, 1924, AJA.

121. Edward R. Stone, "Weekly Reports—Concentration Camps (Reasons and Places)," July 16, 1924, Agregado Militar de Estados Unidos: Informes: Inv. 32 Exp. 090102 (Serie 1924) Leg. 5/6, ff. 67; George M. Russell, "Reorganization of the Mexican Army," August 2, 1924, Agregado Militar de Estados Unidos: Correspondencia: Inv. 31 Exp. 090101 Leg. 1 Serie 1924, ff. 33, both in CDEEUM.

122. No title, May 1924, Agregado Militar de Estados Unidos: Informes Anónimos, Inv. 33 Exp. 090103 Leg.6/12, CDEEUM.

123. Prieto, *Mexican Expedition*, 48, 69, 68.

124. Seely, *Building the American Highway System*, 104.

125. The general manager for the Society of Automotive Engineers wrote, "In designing the United States war trucks the engineers have naturally followed closely the experience gained by Allies, as well as that had on the Mexican border

by our manufactures and Army officers"; Coker F. Clarkson, "Automotives in the Great War," *Scientific American*, January 26, 1918, 54.

126. US soldiers occupying Haiti, Nicaragua, Cuba, Panama, and the Dominican Republic all increasingly partook in more road-building projects between 1916 and 1924. However, communications seldom improved, and violence usually escalated. Military engineering projects also fed the developing construction and automobile industries in the United States, which in 1927 exported more than two thousand concrete mixers, 12,382,000 pounds of road-making machinery, and 380,000 vehicles. According to one author, Latin American expenditures on roads alone were over half a billion dollars in 1927. See John Carter, "This Road Building Era," *North American Review* 228, no. 5 (November 1929): 597.

CHAPTER 8: FLYING MACHINES AS A MEASURE OF MEXICO

1. Henry Woodhouse, president of the Aerial League of America, to President Plutarco Elías Calles, June 8, 1925, Plutarco Elías Calles, exp. 711-W-9, Archivo General de la Nación, Mexico City (hereafter AGN).

2. The cultural revolution was complex and, in many regards, meant for external consumption, but the cultural programs of the era frequently shared a mission to redefine national identity in opposition to the Porfirians. A rejection of European and US influences, at least on the surface, was an important part of that. To this extent, the revolution can be characterized as inward-looking. It was therefore necessary for officials to balance cosmopolitanism with nationalism. Numerous scholars have touched on the complex and at times contradictory nature of the cultural revolution. See, for example, Alan Knight, "Popular Culture and the Revolutionary State in Mexico, 1910–1940," *Hispanic American Historical Review* 7, no. 3 (1994): 393–444; Rick Lopez, "The Noche Mexicana and the Exhibition of Popular Arts: Two Ways of Exalting Indianness," in *The Eagle and the Virgin*, ed. Mary Kay Vaughn and Stephen E. Lewis (Durham, NC: Duke University Press, 2006), 23–42; Justin Castro, *Radio in Revolution: Wireless Technology and State Power in Mexico, 1897–1938* (Lincoln: University of Nebraska Press, 2016).

3. Roberto Fierro discusses flying with traveling air shows in his memoir, *Esta es mi vida* (Mexico City: Los Talleres Gráficos, 1964), 73–76.

4. Automobile clubs played a strikingly similar role in advancing road construction and advocating for auto mobility both in Mexico and elsewhere in Latin America. See Jayson Porter's chapter in this volume; Michael K. Bess, *Routes of Compromise: Building Roads and Shaping the Nation in Mexico, 1917–1952* (Lincoln: University of Nebraska Press, 2017), 7, 22, 73, 85, 136–37; and Brian Freeman, "'Los Hijos de Ford': Mexico in the Automobile Age, 1900–1930," in *Technology and Culture in Twentieth-Century Mexico*, ed. Araceli Tinajero and J. Brian Freeman (Tuscaloosa: The University of Alabama Press, 2013), 225–27. Wendy Waters also talks about the importance of community organizing in bringing roads and automobiles to provincial areas; Waters, "Remapping Identities: Road

Construction and Nation Building in Postrevolutionary Mexico," in Vaughn and Lewis, *Eagle and the Virgin*, 227. For Brazil, see Joel Wolfe, *Autos and Progress: The Brazilian Search for Modernity* (New York: Oxford University Press, 2010), 22, 31, 36, 46–57, 97–98.

5. Thomas Hughes, "Technological Momentum," in *Does Technology Drive History? The Dilemma of Technological Determinism*, ed. Merritt Roe Smith and Leo Marx (Cambridge, MA: MIT Press, 1994), 101–13.

6. Michael Adas, *Machines as the Measure of Men: Science, Technology, and Ideologies of Western Dominance* (Ithaca, NY: Cornell University Press, 1989).

7. Wiliam H. Beezley, *Judas at the Jockey Club* (Lincoln: University of Nebraska, 2004), 6.

8. Paul Vanderwood, *Disorder and Progress: Bandits, Police, and Mexican Development* (Lincoln: University of Nebraska Press, 1981), 87; Michael Matthews, "*De Viaje*: Elite Views of Modernity and the Porfirian Railway Boom," *Mexican Studies/Estudios Mexicanos* 26, no. 2 (2010): 225.

9. Victor Niemeyer, *Revolution at Querétaro: The Mexican Constitutional Convention of 1916–1917* (Austin: University of Texas Press, 1974), 60–71.

10. Joanne Hershfield, "Domestic Technologies: Gender, Technology, and Mexican Housewives, 1930–1950," in Tinajero and Freeman, *Technology and Culture in Twentieth-Century Mexico*, 55–56; Castro, *Radio in Revolution*, chaps. 4–6.

11. Manuel Rúiz Romero, *Grandes vuelos en la aviación mexicana* (Mexico City: Grupo Editorial Aviación, S.A., 1986), 37–57. For Villa's interactions with US film companies, see Margarita de Orellana, *Filming Pancho Villa: How Hollywood Shaped the Mexican Revolution* (New York: Verso, 2009); E. V. Niemeyer, *Revolution at Querétaro: The Mexican Constitutional Convention 1916–1917* (Austin: University of Texas Press, 1974), 101.

12. Mary K. Vaughan and Stephen E. Lewis, introduction to Vaughn and Lewis, *Eagle and the Virgin*, 9.

13. Helen Delpar, "Mexican Culture, 1920–45," in *The Oxford History of Mexico*, ed. Michael Meyer and William Beezley (New York: Oxford University Press, 2000), 543–47.

14. Castro, *Radio in Revolution*, 106–7; José Ramón Buergo Troncoso, "Alas en el Cine Mexicana," *Quauhtli* 2 (2000): 28.

15. Woodhouse to Calles, June 8, 1925; report from the Mexican Aeronautics Association to the Secretary of Foreign Relations, October 28, 1929, exp. III-48-4, Secretaría de Relaciones Exteriores, Archivo Historico "Genaro Estrada," Mexico City (hereafter SRE)

16. Willie Haitt, *The Rarefied Air of the Modern: Airplanes and Technological Modernity in the Andes* (New York: Oxford University Press, 2016), 69–74.

17. For the column Literatura Mexicana, see, for example, *Tohtli: Revista de Aeronautica Militar* 2, no. 1 (1917) to 3, no. 12 (1918), HathiTrust Digital Library, https://babel.hathitrust.org/cgi/pt?id=uiug.30112112405474;view=1up;seq=9.

18. Knight, "Popular Culture and the Revolutionary State," 419–20.

19. Scott Palmer, *Dictatorship of the Air: Aviation Culture and the Fate of Modern Russia* (New York: Cambridge University Press, 2006), 136–54.

20. Another famous example being pioneering Soviet filmmaker Sergei Eisenstein's visit to the country to film *¡Que Viva México!*

21. Castro, *Revolution in Radio*, 73–106, 128–50.

22. Manuel Ruiz Romero, "Plutarco Elías Calles y la aviación Mexicana," in *Quauhtli* 1 (1999): 49–51.

23. Romero, "Plutarcho Elías Calles y la aviación Mexicana," 49–51; Fierro, *Esta es mi vida*, 50–56, 78–81; Evelyn Hu-DeHart, "Peasant Rebellion in the Northwest: The Yaqui Indians of Sonora, 1740–1976," in *Riot, Rebellion, and Revolution: Rural Social Conflict in Mexico*, ed. Friedrich Katz (Princeton, NJ: Princeton University Press, 1988), 167–68.

24. José Villela Gomez, *Breve historia de aviación mexicana* (Mexico City: Complejo Editorial Mexicano, 1971), 209–19.

25. Villela Gomez, *Breve historia de aviación mexicana*, 216–17.

26. See Marcela Saldaña Solís's chapter in this volume; Barbara Hibino, "Cervecería Cuauhtémoc: A Case Study of Technological and Industrial Development in Mexico," *Mexican Studies/Estudios Mexicanos* 8, no. 1 (1992): 29.

27. See James Garza's chapter in this volume.

28. Villela Gomez, *Breve historia de aviación mexicana*, 220–23; Juan F. Acárate to Admiral A. B. Vosseller, assistant director of the development planning department at Lockheed Aircraft Corporation, February 24, 1958, Fideicomiso Archivos Plutarco Elías Calles y Fernando Torreblanca, Associaciones, Empresas e Industrie, 1915–1967 exp. Compañía Constructora de Aeroplanos, S.A., Archivo Abelardo L. Rodríguez, Mexico City; Juan F. Azcárate to President Pascual Ortiz Rubio, November 22, 1930, Fondo Ortíz Rubio exp. 179 (1930) 14109, AGN.

29. Fierro, *Esta es mi vida*, 47–48.

30. Departamento de Aeronautica Civil Sección de Estadística y Proyectos, "Servicio Federal Aereo," report, April 16, 1929, Legajo L-E 209 vol. 2, Sexta Conferencia Internacional Americana, SRE.

31. Report no. 42, June 8, 1933, Fondo Secretaría de Communicaciónes y Tranportes (SCT)—Dirreción General de Aeronautica Civil (DGAC), caja 149 exp. 35/016/1–3, AGN.

32. Katherine Bliss, "For the Health of the Nation: Gender and the Cultural Politics of Social Hygiene in Revolutionary Mexico," in Vaughn and Lewis, *Eagle and the Virgin*, 196–97; Gabriela Soto Laveaga, "Bringing the Revolution to Medical Schools: Social Service and a Rural Health Emphasis in 1930s Mexico," *Mexican Studies/Estudios Mexicanos* 29, no. 2 (2013): 401.

33. Report from the head of the technical section of the Department of Civil Aeronautics to the head of the Department of Civil Aeronautics, June 25, 1929, SCT-DGAC, caja 148 exp. 1X/061.10/29–370128; Physical for pilot-aviator Juan

Jenkins, December 11, 1929, SCT-DGAC, caja 148 exp. 1X/061.4–370126, both in AGN.

34. Report from the head of the technical section of the Department of Civil Aeronautics to the head of the Department of Civil Aeronautics, July 9, 1930, SCT-DGAC, caja 55 exp. 1X/014/2–29803, AGN.

35. Randall Hansis, "The Political Strategy of Military Reform: Álvaro Obregón and Revolutionary Mexico, 1920–1924," *The Americas* 26, no. 2 (1979): 203.

36. Stephen Haber, *Industry and Underdevelopment: The Industrialization of Mexico, 1890–1940* (Stanford, CA: Stanford University Press, 1989), 156.

37. "El Presidente inaugurará la exposición aeronáutica," *El Universal,* December 6, 1929; Manuel Ruiz Romero, *Aeropuertos: Historia de la construcción, operación y administración aeropuertaria en México* (Mexico City: Aeropuertos y Servicios Auxiliares, 2003), 45–48.

38. "El Señor Presidente hizo un viaje en un trimotor," *El Universal,* January 12, 1929.

39. Rafael R. Esparza, *Historia de las comunicaciones y los transportes en México: Aviación* (Mexico City: Secretaría de Communicaciónes y Transportes, 1987), 207; Ruiz Romero, *La aviación civil en México* (Mexico City: Universidad Nacional Autónoma de México, 1999), 128.

40. Manuel Ruiz Romero, *Diccionario biografico aeronautico de Mexico* (Mexico City: El Universal, 2002), 207.

41. President Plutarco Elías Calles to the Secretariat of Communications, Fondo Obregón-Calles, "Acuerdos Presidenciales," 665, 666, 669, and 632–1928, AGN; Waters, "Remapping Identities," 221.

42. Jürgen Buchenau, *Plutarco Elías Calles and the Mexican Revolution* (Lanham, MD: Rowman and Littlefield, 2007), 115, 121; Waters, "Remapping Identities," 221–42.

43. Obregón-Calles, Acuerdo 665–1928.

44. Enrique Cárdenas de la Peña, *Historia de las comunicaciones y los transportes en México: El correo* (Mexico City: Secretaría de Communicaciónes y Transportes, 1987), 202, 205.

45. Private companies often built their own airstrips, and the Compañía Mexicana de Aviación helped fund construction of the Central Airport in Mexico City. See Romero, *Aeropuertos,* 46–47; see also Memorandum Jucio de Amparo—Aeropuerto "Mazatlan," (n.d.), Fondo Abelardo Rodríguez exp. 522/110, AGN.

46. Abbott Lawrence Rotch, qtd. in Woodhouse to Calles, June 8, 1925.

47. Calles-Obregón, Acuerdo 665–1928; Cárdenas de la Peña, *El Correo,* 202–6; Waters, "Remapping Identities," 224; Lars Denicke, "Fifty Years' Progress in Five Brasilia: Modernization, Globalism, and the Geopolitics of Flight," in *Entangled Geographies: Empire and Technopolitics in the Cold War,* ed. Gabrielle Hecht (Cambridge, MA: MIT Press, 2011), 191–92.

48. The Civil Aeronautics Act, Ch. VI, Articles 43, 44, 47, 73, and 74, SCT-DGAC, caja 55 exp. 1X/011/2–19794 (1929), AGN.

49. General Communications Law, Vol. 4, Ch. III, Article 440 (Mexico City: Secretaría de Comunicaciones y Obras Públicas, 1932), 152.

50. Buchenau, *Plutarco Elías Calles*, 119.

51. Tzvi Medin, *Ideología y prazis politica de Lázaro Cárdenas* (Mexico City: Siglo Vientiuno Editores, 1987), 74–94.

52. James Wilke, *The Mexican Revolution: Federal Expenditure and Social Change since 1910*, 2nd ed. (Berkeley: University of California Press, 1967). See also Ariel Jose Contreras, *México 1940: Industrialización y crisis política* (Mexico City: Siglo Vientiuno editors, 1977).

53. Elfego Castaneda to President Cárdenas, September 21, 1938, Fondo LCR, caja 944 exp. 552.3/65, AGN.

54. These letters can be found at LCR, caja 944 exp. 552.3/1 through 552.3/60 and caja 945 exp. 552.3/61 through 552.3/72, AGN.

55. Moisés Guzmán Pérez, *Pinocho: Una página en la historia de la Aviación Mexicana* (Morelia: Universidad Michoacana de San Nicolás de Hidalgo, 1998), 18–35, 41–44.

56. José D. Macías to President Cárdenas, June 30, 1938, LCR, caja 598 exp. 511.1/26; Leonides Garcia Agular to President Ávila Camacho July 25, 1942, Ávila Camacho, caja 852, exp. 552.6/11; Dr. Régulo Torpey to President Ávila Camacho, October 13, 1945, exp. 552.6/20; Palizada Ejidal Commission to President Ruíz Cortines, June 1, 1956, Ruíz Cortines, caja 641, exp. 511.1/17; San Augustín de las Juntas Ejidal Commission to President Ruíz Cortines, May 6, 1955, exp. 511.1/32, all in AGN.

57. Adolfo Gilly, *La revolución interrumpida*, 12th ed. (Mexico City: Ediciones El Caballito, 1971).

58. Maria Vargas-Lobsinger, *La Comarca Lagunera de la Revolución de las haciendas, 1910–1940* (Mexico City: UNAM Instituto Nacional de Estudios Históricos de la Revolución Mexicana, 1999), 158–75; Jocelyn Olcott, *Revolutionary Women in Postrevolutionary Mexico* (Durham, NC: Duke University Press, 2005), 126–29. See also Mikael Wolfe's *Watering the Revolution: An Environmental and Technological History of Agrarian Reform in Mexico* (Durham, NC: Duke University Press, 2017), esp. chap. 3, 97–127.

59. Coahuila State Governor Ignacio Cepeda Dávila to the president, telegrams, March 26, 1946, Ávila Camacho, caja 548 exp. 511.6/7; El Fresno Ejidal Commission to the president, telegram, July 19, 1945, Ávila Camacho, caja 548 exp. 511.6/7, both in AGN.

60. Hughes, "Technological Momentum," 107–13.

61. Manuel Ávila Camacho, *Mensaje a la nación y otros discursos* (Mexico City: Secretaría de Governación, 1943), 34, Ávila Camacho, caja 277 exp. 161.1/81, AGN; Monica Rankin, "Mexico: Industrialization through Unity," in *Latin Amer-*

ica during World War II, ed. Thomas Leonard and John Bratzel (Lanham, MD: Rowman & Littlefield, 2006), 30.

62. Dan Hagedorn, *Conquistadores of the Sky: A History of Aviation in Latin America* (Gainesville: University Press of Florida, 2008), 327–32, 407–9; Stephen Niblo, *Mexico in the 1940s: Modernity, Politics, and Corruption* (Wilmington, DE: Scholarly Resources, 2009), 117–18; T. E. Braniff to President Ávila Camacho, July 23, 1945, Ávila Camacho, caja 548, 511.6/42, AGN.

63. Elfego Cabrera to President Ávila Camacho, July 29, 1941, Ávila Camacho, caja 397 exp. 432/234, AGN.

64. Ávila Camacho, *Mensaje a la nación y otros discursos*, 34; Rankin, "Mexico," 30; Hagedorn, *Conquistadors of the Sky*, 399–400; Governor Magdaleno Aguilar to President Ávila Camacho, October 21, 1943, Ávila Camacho, caja 850 exp. 552/5, AGN.

65. Hagedorn, *Conquistadors of the Sky*, 411–17; Secretary of Communications Maximino Ávila Camacho to President Ávila Camacho, April 26, 1943, Ávila Camacho, caja 825 exp. 550/23, AGN; William Tudor, "Flight of the Eagles: The Mexican Expeditionary Air Force *Escuadrón 201* in World War II" (PhD diss., Texas Christian University, 1997), 289.

66. Julie Moreno, *Yankee Don't Go Home! Mexican Nationalism, American Business Culture, and the Shaping of Modern Mexico, 1920–1950* (Chapel Hill: University of North Carolina Press, 2003), 47; Dina Berger, *The Development of Mexico's Tourism Industry: Pyramids by Day, Martinis by Night* (New York: Palgrave Macmillan, 2006), 5, 72–73, 76; Francisco Ochoa, head of publicity, Aerovias Braniff, to Secretary Jesús González Gallo, February 18, 1946, Ávila Camacho, caja 549 exp. 511.6/42; Compañía Mexicana de Aviación tourist brochure, circa 1945, caja 548 exp. 511/6/11; Secretary of Communications Maximino Ávila Camacho to President Ávila Camacho, July 31, 1943, caja 549 exp. 511.6/30, all in AGN.

67. Secretaría Comunicaciones y Obras Publicas Departamento de Aeronautica, Civil Map and Legend of Civil Air Concessions, Concessiones, Process, and Operations Permits, Ávila Camacho, caja 1009 exp. 606.3/237, AGN; José Villela Gomez, *Breve historia de la aviación en México* (Mexico City, 1971), 322.

68. Hughes, "Technological Momentum in History," 111.

CHAPTER 9: A SOCIAL HISTORY OF URBAN EXPERTISE

1. For a discussion of urban modernity across the Global South, see Jennifer Robinson, *Ordinary Cities: Between Modernity and Development* (London: Routledge, 2006).

2. Examples include Miguel Angel Centeno and Patricio Silva, *The Politics of Expertise in Latin America* (Houndmills, UK: Palgrave Macmillan, 1998); Sarah L. Babb, *Managing Mexico: Economists from Nationalism to Neoliberalism* (Princeton, NJ: Princeton University Press, 2004); and J. Justin Castro, *Apostle of Progress:*

Modesto C. Rolland, Global Progressivism, and the Engineering of Revolutionary Mexico (Lincoln: University of Nebraska Press, 2019).

3. Urban and technology histories tend to marginalize questions of popular politics. See, for example, Claudia Agostoni, *Monuments of Progress: Modernization and Public Health in Mexico City, 1976–1910* (Calgary: University of Calgary Press, 2003); Manuel Perló Cohen, *El paradigm porfiriano: Historia del desagüe del valle de México* (Mexico City: Porrúa, 1998); and Germán Mejía Pavony, *Los años del cambio: Historia urbana de Bogotá, 1820–1910* (Bogotá: Universidad Javeriana, 2000); and Araceli Tinajero and J. Brian Freeman, eds., *Technology and Culture in Twentieth-Century Mexico* (Tuscaloosa: University of Alabama Press, 2013). There are some exceptions within Latin American historiography, including Mark Healey, *The Ruins of the New Argentina: Peronism and the Remaking of San Juan after the 1944 Earthquake* (Durham, NC: Duke University Press, 2011); and Sarah Hines, "The Power and Ethics of Vernacular Modernism: The Misicuni Dam Project in Cochabamba, Bolivia, 1944–2017," *Hispanic American Historical Review* 98, no. 2 (May 2018): 223–56.

4. See Diana J. Montaño, *Electrifying Mexico: Technology and the Transformation of a Modern City* (Austin: University of Texas Press, 2021).

5. Edward Beatty, *Technology and the Search for Progress in Modern Mexico* (Oakland: University of California Press, 2015), 195–96.

6. For more on the drainage, see James Garza's chapter in this volume.

7. E. A. Martínez Miranda and María de la Paz Ramos Lara, "La física y la formación de los ingenieros mexicanos que colaboraron en el magno proyecto hidroeléctrico Necaxa," *Revista Mexicana de Física* 51, no. 1 (June 2005): 37–44; and William L. Hooper, "The Necaxa Development of the Mexican Light and Power Company," *Proceedings of the American Institute of Electric Engineers* 30, no. 1 (1930): 45–50.

8. Matthew Vitz, *A City on a Lake: Urban Political Ecology and the Growth of Mexico City* (Durham, NC: Duke University Press, 2018), 37–38.

9. Maria Kaika, *City of Flows: Modernity, Nature, and the City* (New York: Routledge, 2012), 57–58.

10. Kaika, *City of Flows*, 35.

11. Matthew Vitz, "Urbanization and the Environment in Mexico since 1521," *Oxford Research Encyclopedia* (July 2016), DOI: 10.1093/acrefore/9780199366439.013.323.

12. Gilbert M. Joseph and Alan Wells, "Modernizing Visions, 'Chilango' Blueprints and Provincial Growing Pains: Mérida at the Turn of the Century," *Mexican Studies/Estudios Mexicanos* 8, no. 2 (Summer 1992): 167–215.

13. Christina Jimenez, "Popular Organizing for Public Services: Residents Modernize Morelia, Mexico, 1880–1920," *Journal of Urban History* 30, no. 4 (May 2004): 507. See also Christina Jimenez, *Making an Urban Public: Popular Claims to the City in Mexico, 1879–1932* (Pittsburgh: University of Pittsburg Press, 2019).

14. Vitz, *City on a Lake*, 48–49. For petitions for lighting couched in discourses of security and aesthetics, see Fracisco Javier Delgado Aguilar, "La demanda de alumbrado," in *Ciudades mexicanas del siglo XX, siete estudios históricos*, ed. Carlos Lira Vásquez and Ariel Rodríguez Kuri (Mexico City: Colegio de México, 2009), 236–37.

15. For a similar argument, see Ariel Rodríguez Kuri, *La historia del desasosiego: La revolución en la ciudad de México* (Mexico City: Colegio de México, 2010).

16. This is an inversion of the technocratic ideal as described in Tania Murray Li, *The Will to Improve: Governmentality, Development and the Practice of Politics* (Durham, NC: Duke University Press, 2007).

17. Anna Rose Alexander, *City on Fire: Technology, Social Change and the Hazards of Progress in Mexico City, 1860–1910* (Pittsburgh: University of Pittsburgh Press, 2016); and Anna Rose Alexander, "Incendiary Legislation: Fire Risk and Protection in Porfirian Puebla," *Mexican Studies/Estudios Mexicanos* 29, no. 1 (Winter 2013): 175–99.

18. Claudia Agostoni, "Material Culture, Public Health, and the Technologies of Hygiene in Modern Mexico," in Tinajero and Freeman, *Technology and Culture*, 35.

19. Rafael Torres Sánchez, *Revolución y vida cotidiana: Guadalajara, 1914–1934* (Sinaloa, Mexico: Universidad Autónoma de Sinaloa, 2001), 207–50.

20. Heron Proal, qtd. in Manuel Castells, *City and the Grassroots: A Cross-Cultural Theory of Urban Social Movements* (Berkeley: University of California Press, 1983), 47.

21. For the theory of the right to city that moves beyond the original woolly formulation by Henri Lefebvre, see Peter Marcuse, "From Critical Urban Theory to the Right to the City," *City* 13 (2009): 185–97; and David Harvey, "The Right to the City," *New Left Review* 53 (September–October 2008): 23–40.

22. Vitz, *City on a Lake*, 92.

23. Lynda Klich, "*Estridentismo*'s Technologies: Modernity's Efficient Agents in Postrevolutionary Mexico," 264–70; and J. Brian Freeman, "Los Hijos de Ford: Mexico in the Automobile Age, 1910–1930," 222–24, both in Tinajero and Freeman, *Technology and Culture*.

24. See Vitz, *City on a Lake*, 95–96; and Ariel Rodriguez Kuri, "Desabasto de agua y violencia política: El motín del 30 de noviembre de 1922 en la ciudad de México; Economia moral y cultura política," in *Formas de descontento y movimientos sociales Siglos XIX y XX*, ed. José Ronzón and Carmen Valdes (Mexico City: UAM Azcapotzalco, 2000); and Stephen Bocking, *Nature's Experts: Science, Politics, and the Environment* (New Brunswick, NJ: Rutgers University Press, 2008). For more on Rolland, see Castro, *Apostle of Progress*.

25. Vitz, *City on a Lake*, 99.

26. Vicente Lombardo Toledano, "La supresión del municipio libre en el Distrito Federal," *Planificación* 1, no. 9 (May 1928): 17–24.

27. Vitz, *City on a Lake*,

28. See Blanca Estela Suárez Cortes and Diana Birrichaga Gardida, *Dos estudios sobre usos del agua en México, siglos XIX y XX* (Jiutepec, Morelos: CIESAS, 1997); and Vivienne Bennett, *The Politics of Water: Urban Protest, Gender, and Power in Monterrey, Mexico* (Pittsburgh: University of Pittsburgh Press, 1996).

29. Sarah Selvidge, "Carlos Lazo and the Politics of Planning in Postrevolutionary Mexico," unpublished manuscript, 2019.

30. Fundación ICA, *ICA: Hacemos realidad, grandes ideas* (Mexico City: ICA, 1997).

31. See Diane Davis, *Urban Leviathan: Mexico City in the Twentieth Century* (Philadelphia: Temple University Press, 1994), 92–97; and Soledad Loaeza, *Clases medias en la coyuntura actual* (Mexico City: Colegio de México, 1990).

32. See Oscar Lewis, *The Children of Sánchez: Autobiography of a Mexican Family* (New York: Random House, 1961). Also see Ariel Rodríguez Kuri, "Simpatía por el diablo: Miradas académicas a la ciudad de México, 1900–1970," in *Los ultimos cien, los próximos cien*, ed. Ariel Rodríguez Kuri and Sergio Tamayo Flores-Alatorre (Mexico City: UNAM, 2004), 45–68.

33. Luis Aboites Aguilar, "The Illusion of National Power: Water Infrastructure in Mexican Cities, 1930–1990," in *A Land between Waters: Environmental Histories of Modern Mexico*, ed. Christopher R. Boyer (Tucson: University of Arizona Press, 2014), 231–32; and Bennett, *Politics of Water*.

34. Aboites Aguilar, "Illusion of National Power," 230–31.

35. Sergio Miranda Pacheco, "El Frankenstein urbano: Ecólogos, urbanistas, e ingenieros frente a la crisis hidrológica de la ciudad de México a mitad del siglo XX," *Historia Ambiental, Latinoamericana y Caribeña* 10, no. 2 (2020): 172.

36. Vitz, *City on a Lake*, 228–29; Manuel Perló Cohen, *Historia de las obras, planes y problemas hidráulicos en el Distrito Federal* (Mexico City: Universidad Nacional Autónoma de México, 1989).

37. Dean Mohammed Chahim, "Flood Control Politics: Engineering, Urban Growth, and Disaster in Mexico City" (PhD diss., Stanford University, 2021), 89–112.

38. Robert Ríos Elizondo, *Memorias de las obras del drenaje profundo del Distrito Federal* (Mexico City: DDF, 1975). In 1973 the city released the striking television video "La obra oculta" to explain the technical marvels of the underground work. https://www.youtube.com/watch?v=RnF7O7SPeeA.

39. Martha Schteingart, *Los productores del espacio habitable: Estado, empresa y sociedad en la ciudad de México* (Mexico City: Colegio de México, 2001), 139–40.

40. This included parts of Ex-Hipódromo de Peralvillo, Obregón's failed worker housing initiative. See George F. Flaherty, *Hotel Mexico: Dwelling on the 68 Movement* (Oakland: University of California Press, 2017), 205–6; and "Anuncio de television sobre la venta de apartmanetos en Conjunto Urbano Presidente Adolfo López Mateos, Ciudad de México, 1964," in Manuel de Sevilla, dir., *Obras para México*, video, 7:31.

41. Anana Roy, "Transnational Trespassings," in *Urban Informality: Transnational Perspectives from the Middle East, Latin America, and South Asia*, ed. Ananya Roy and Nezar Aslsayyad (Lanham, MD: Lexington Books, 2004).

42. Mauricio Gómez Mayorga, *Qué hacer por la ciudad de México* (Mexico City: Costa-Amic, 1957), 45, 49. All translations from the Spanish are mine.

43. Adolfo Zamora, *El problema de la habitación en la ciudad de México* (Mexico City: BANHUOP, 1952), 15, 141.

44. Pablo Landa, "Overflowing Architecture: Home, Neighborhood, and Nation in Mexico's Modern Experience" (PhD diss., Princeton University, 2015), chap. 1.

45. Landa, "Overflowing Architecture," chap. 2.

46. Jan Bazant Sánchez, *Autoconstrucción de vivienda popular* (Mexico City: Trillas, 1991), 63–128; and Jorge Legorreta, *La autoconstrucción de vivienda en México, el caso de las ciudades petroleras* (Mexico: CECODES, 1984), 77–78.

47. For communal labor in home construction, see Carlos Vélez Ibañez, *Rituals of Marginality, 1969–1974: Politics, Process, and Cultural Change in Central Urban Mexico* (Berkeley: University of California Press, 1991), 69–70; and Bennett, *Politics of Water*.

48. See Michael C. Ennis-McMillan, "Women, Equity, and Household Water Management in the Valley of Mexico," in *Opposing Currents: The Politics of Water and Gender in Latin America*, ed. Vivienne Bennett, Sonia Dávila Poblete, and María Nieves Rico (Pittsburgh: University of Pittsburgh Press, 2005); and Bennett, *Politics of Water*, 77.

49. See Vitz, *City on a Lake*, chap. 6.

50. James Holston, *Insurgent Citizenship: Disjunctions of Democracy and Modernity in Brazil* (Princeton, NJ: Princeton University Press, 2009), 23.

51. Vélez Ibañez, *Rituals of Marginality*, 97, 124–25; and Kenneth Maffitt, "Nueva política social, Viejo contrato social: Políticas de vivienda y protesta urbana en la periferia de la ciudad de México, 1960s–1980s," *Historia* 47, no. 1 (January–July 2014): 121–23.

52. Maffitt, "Nueva política," 122–23.

53. For an analysis of social Maoism after 1968, see Michael Soldatenko, "The Various Lives of Mexican Maoism: Política Popular, a Mexican Social Maoist Praxis," in *México beyond 1968: Revolutionaries, Radicals, and Repression during the Global Sixties and Subversive Seventies*, ed. Jaime M. Pensado and Enrique C. Ochoa (Tucson: University of Arizona Press, 2018), 175–94.

54. Juan Manuel Ramírez Saíz, *Carácter y contradicciones de la ley general de asentamientos humanos* (Mexico City: UNAM, 1983), 21; Mike Douglas and John Friedman, *Cities for Citizens: Planning and the Rise of Civil Society in a Global Age* (New York: Wiley, 1998); and Daniel Immerwahr, *Thinking Small: The United States and the Lure of Community Development* (Cambridge, MA: Harvard University Press, 2017).

55. Ramirez Saiz, *Carácter y contradicciones*, 111–19.

56. Jan Bazant Sánchez, *Autoconstrucción*, 40–62.

57. Bennett, *Politics of Water*, 105.

58. Juan Manuel Ramírez Sáiz, *El movimiento urbano popular en México* (Mexico City: Siglo XXI, 1999), 93–94.

59. For reports on pollution, see Consultivo Técnico, 287, 2442, Archivo Histórico del Agua, Mexico City; and MMH, SEDUE, 20.04.01.00 box 33, file 2, Archivo General de la Nación, Mexico City (hereafter AGN).

60. Ramírez Sáiz, *El movimiento urbano popular*, 94–103.

61. See, for example, Priscilla Connolly, "The Go-Between: CENVI, a Habitat NGO in Mexico City," *Environment and Urbanization* 5, no. 1 (April 1993): 68–90.

62. Connolly, "Go-Between," 154–63; and Pedro Moctezuma, "La Coordinadora Nacional del Movimiento Urbano Popular en el Valle de México," in *Los movimientos sociales en el valle de México*, ed. Jorge Alonso (Mexico City: CIESAS, 1986), 209.

63. Ricardo Hernández, *La Coordinadora Nacional del Movimiento Urbano Popular, CONAMUP: Su historia, 1980–1986* (Mexico City: Praxis Gráfica, 1987), 92, 110; Pedro Moctezuma Barragán, *Despertares: Comunidad y organización urbano-popular en México* (Mexico City: Universidad Iberoamericana, 1999), 135–38; and "Mexico's Urban Popular Movements: A Conversation with Pedro Moctezuma," *Environment and Urbanization* 2, no. 1 (April 1990): 40–42.

64. Hernández, *La Coordinadora*, 92; and Moctezuma, "La Coordinadora," 211.

65. Keith Pezzoli, *Human Settlements and Planning for Ecological Sustainability: The Case of Mexico City* (Cambridge, MA: MIT Press, 2000).

66. The classic reference for the rise of civil society is Carlos Monsivais, *Entrada libre: Crónicas de la sociedad que se organiza* (Mexico City: ERA, 1994). See also Sergio Tamayo, "Democracia en la ciudad: Desde los barrios," in *Política y movimientos sociales en la ciudad de México*, ed. Alfonso X Iracheta and Albero Villar Calvo (Mexico City: Plaza y Valdés, 1998), 97–126.

67. Diane Davis, "Reverberations: Mexico City's 1985 Earthquake and the Transformation of the Capital," in *The Resilient City: How Modern Cities Recover from Disaster*, ed. Lawrence J. Vale and Thomas J. Campanella (New York: Oxford University Press, 2005), 264–65; and Elena Poniatowska, *Nothing, Nobody: The Voices of the Mexico City Earthquake* (Philadelphia: Temple University Press, 1995).

68. Louise Walker, *Waking from the Dream: Mexico's Middle Classes after 1968* (Stanford, CA: Stanford University Press, 2014), 176–77.

69. Poniatowska, *Nothing, Nobody*, 93.

70. Francisco Nuñez de la Peña and Jesus Orozco, *Terremoto: Una version corregida* (Mexico City: Iteso, 1988), 156.

71. Poniatowska, *Nothing, Nobody*, 8, 32; and Nuñez and Orozco, *Terremoto*, 99–101.

72. Davis, "Reveberations," 261.

73. Poniatowska, *Nothing, Nobody*, 255.

74. Walker, *Waking from the Dream*, 181; and Ruben Gallo, "Tlatelolco," in *Noir Urbanisms: Dystopic Images of the Modern City*, ed. Gyan Prakash (Princeton, NJ: Princeton University Press, 2011), 65.

75. Poniatowska, *Nothing, Nobody*, 84.

76. Poniatowska, *Nothing, Nobody*, 84.

77. Leslie Serna and Cristina Pacheco, *Aqui nos quedaremos: Testimonios de la coordinadora unica de damnificados* (Mexico City: UIA, 1995), 60–61.

78. Poniatowska, *Nothing, Nobody*, 264.

79. José Luis Mecatl, Marco A. Michel, and Alicia Ziccardi, *Casa a los Damnificados: Dos años de política habitacional en la reconstrucción de la ciudad de México, 1985–1987* (Mexico City: UNAM, 1988), 22.

80. On the political-ideological differences between CUD and CONAMUP, see Raúl Bautista González, *Movimiento urbano popular: Bitácora de lucha, 1968–2011* (Mexico City: Casa y Ciudad, 2015), 24–25.

81. Davis, "Reverberations," 268–70; and Walker, *Waking from the Dream*, 181.

82. "Relación de la sesión jueves 7," Fondo Miguel de la Madrid Hurtado, Universidades, 32.06.00.00, caja 1, AGN.

83. Cuauhtémoc Abarca, qtd. in Luis Mecatel et al., *Casa a los damnificados*, 62–63.

84. Mecatel et al., *Casa a los damnificados*, 64.

85. Serna and Pacheco, *Aqui nos quedaremos*, 50–53; and "Proyecto para una vecindad ecológica en el centro histórico," Fondo Miguel de la Madrid Hurtado, Universidades, 32.06.00.00, caja 1, AGN.

86. Serna and Pacheco, *Aqui nos quedaremos*, 102–3, 115–16; and Secretaria de Desarrollo Urbano y Ecología, *Alternativas de vivienda en barrios populares* (Mexico City: UAM-Xochimilco, 1988).

87. Miguel Armas, qtd. in Serna and Pacheco, *Aqui nos quedaremos*, 65.

88. Pedro Moctezuma Barragán, "Community-Based Organizations and Participatory Planning in Southeastern Mexico City," *Environment and Urbanization* 13, no. 2 (2001): 117–33.

89. Citizen activism has defeated several projects in Mexico City, including the original plan for a Xochimilco tourist zone, the Reforma shopping corridor, and the Texcoco airport.

SELECTED BIBLIOGRAPHY

Aboites, Luis, and Engracia Loyo. "La construcción del nuevo Estado, 1920–1943." In *Nueva Historia General de México*, edited by Erick Velázquez García et al. Mexico City: El Colegio de México, 2010.

Adas, Michael. *Machines as the Measure of Men: Science, Technology, and Ideologies of Western Dominance*. Ithaca, NY: Cornell University Press, 1989.

Agostoni, Claudia. *Monuments of Progress: Modernization and Public Health in Mexico City, 1876–1910*. Calgary: University of Calgary Press, 2003.

Alexander. Anna Rose. *City on Fire: Technology, Social Change, and the Hazards of Mexican Progress, 1860–1910*. Pittsburgh: University of Pittsburgh Press, 2016.

Ávila Galinzoga, Jesús, ed. *La educación técnica en México desde la Independencia, 1810–2010*. Vol. 2. Mexico City: Instituto Politécnico Nacional, 2011.

Bakewell, P. J. *Silver Mining and Society in Colonial Mexico: Zacatecas, 1546–1700*. Cambridge: Cambridge University Press, 1972.

Bazant, Mílada. "La enseñanza y la práctica de la ingeniería durante el Porfiriato." *Historia Mexicana* 33, no. 3 (January–March 1984): 254–97.

Bazant, Mílada. "Estudiantes mexicanos en el extranjero: El caso de los hermanos Urquidi." *Historia Mexicana* 36, no. 4 (April–June 1987): 739–58.

Bazant, Mílada. *Historia de la educación durante el Porfiriato*. Mexico City: El Colegio de México, Centro de Estudios Históricos, 2006.

Beatty, Edward. *Technology and the Search for Progress in Modern Mexico*. Oakland: University of California Press, 2015.

Bertsou, Eri, and Daniele Caramani. *The Technocratic Challenge to Democracy.* New York: Routledge, 2020.

Bigelow, Allison Margaret. *Mining Language: Racial Thinking, Indigenous Knowledge and Colonial Metallurgy in the Early Modern Iberian World.* Williamsburg, VA: Omohundro Institute of Early American History and Culture; and Chapel Hill: University of North Carolina Press, 2020.

Boyer, Christopher R. *A Land between Waters: Environmental Histories of Modern Mexico.* Tucson: University of Arizona Press, 2014.

Buchenau, Jürgen. *The Last Caudillo: Álvaro Obregón and the Mexican Revolution.* Malden, MA: Wiley-Blackwell, 2011.

Buckley, Eve E. *Technocrats and the Politics of Drought and Development in Twentieth-Century Brazil.* Chapel Hill: University of North Carolina Press, 2017.

Bunker, Steven B. *Creating Mexican Consumer Culture in the Age of Porfirio Díaz.* Albuquerque: University of New Mexico Press, 2012.

Candiani, Vera S. *Dreaming of Dry Land: Environmental Transformation in Colonial Mexico City.* Stanford, CA: Stanford University Press, 2014.

Casas, Rosalba. *El estado y la política de la ciencia en México.* Mexico City: IISUNAM, 1985.

Castro, J. Justin. *Apostle of Progress: Modesto C. Rolland, Global Progressivism, and the Engineering of Revolutionary Mexico.* Lincoln: University of Nebraska Press, 2019.

Chasteen, Sandra B., and Timothy W. Lorek, eds. *Itineraries of Expertise: Science, Technology, and Environment in Latin America's Long Cold War.* Pittsburgh: University of Pittsburgh Press, 2020.

Clendenen, Clarence C. *The United States and Pancho Villa: A Study in Unconventional Diplomacy.* Ithaca, NY: Cornell University Press, 1961.

Connolly, Priscilla. *El contratista de don Porfirio: Obras públicas, deuda y desarrollo desigual.* Mexico City: Fondo de Cultura Económica, 1997.

Craib, Raymond B. *Cartographic Mexico: A History of State Fixations and Fugitive Landscapes.* Durham, NC: Duke University Press, 2004.

Dalton, David. *Race, Technology, and the Body in Post-Revolutionary Mexico.* Gainesville: University Press of Florida, 2018.

Davis, Diane. *Urban Leviathan: Mexico City in the Twentieth Century.* Philadelphia: Temple University Press, 1994.

De la Torre, Federico. *La ingeniería en Jalisco en el siglo XIX.* Guadalajara: Universidad de Guadalajara, 2000.

Derickson, Alan. *Black Lung: Anatomy of a Public Health Disaster.* Ithaca, NY: Cornell University Press, 2014.

Domínguez Martínez, Raúl. *La ingeniería civil en México, 1900–1940: Análisis histórico de los factores de su desarrollo.* Mexico City: Universidad Autónoma Nacional de México, 2013.

England, Shawn. "'Mexicans Are Good Flyers': Militarized Airpower, Aviation Idols, and Aviation Diplomacy in Revolutionary Mexico." *Canadian Journal of Latin American and Caribbean Studies* 40 no. 3 (2015): 411–28.

Escamilla González, Francisco Omar, ed. *200 años del Palacio de Minería, su historia a partir de fuentes documentales.* Mexico City: UNAM-Facultad de Ingeniería, 2013.

Fundadores del Instituto de Ingeniería: Inteligencia y pasión. Mexico City: UNAM-Instituto de Ingeniería, 2014.

Garner, Paul. *British Lions and Mexican Eagles: Business, Politics and Empire in the Career of Weetman Pearson in Mexico 1889–1919.* Stanford, CA: Stanford University Press, 2011.

Garner, Paul. *Porfirio Díaz.* London: Routledge, 2001.

Gomez, José Villela. *Breve historia de aviación Mexicana.* Mexico City, 1971.

Gomez, Rocio. *Silver Veins, Dusty Lungs: Mining, Water, and Public Health in Zacatecas, 1835–1945.* Lincoln: University of Nebraska Press, 2020.

Gómez-Galvarriato, Aurora. *Industry & Revolution: Social and Economic Change in the Prizaba Valley, Mexico.* Cambridge, MA: Harvard University Press, 2013.

Haber, Steven. *Industry and Underdevelopment: The Industrialization of Mexico, 1890–1940.* Stanford, CA: Stanford University Press, 1989.

Hagedorn, Dan. *Conquistadors of the Sky: A History of Aviation in Latin America.* Gainesville: University Press of Florida, 2008.

Harris, Charles H., III, and Louis R. Sadler. *The Great Call-Up: The Guard, the Border, and the Mexican Revolution.* Norman: University of Oklahoma Press, 2016.

Hewitt de Alcántara, Cynthia. *Modernizing Mexican Agriculture: Socioeconomic Implications of Technological Change, 1940–1970.* Geneva: United Nations Research Institute for Social Development, 1976.

Holston, James. *Insurgent Citizenship: Disjunctions of Democracy and Modernity in Brazil.* Princeton, NJ: Princeton University Press, 2009.

Jimenez, Christina. *Making an Urban Public: Popular Claims to the City in Mexico, 1879–1932*, Pittsburgh: University of Pittsburg Press, 2019.

Kaika, Maria. *City of Flows: Modernity, Nature, and the City.* New York: Routledge, 2012.

Katz, Friedrich. *The Life & Times of Pancho Villa.* Stanford, CA: Stanford University Press, 1998.

Katz, Friedrich. *The Secret War in Mexico: Europe, the United States, and the Mexican Revolution.* Chicago: University of Chicago Press, 1981.

Knight, Alan *U.S.-Mexican Relations, 1910–1940: An Interpretation.* San Diego: Center of U.S.-Mexican Studies, University of California, 1987.

Latour, Bruno. *Reassembling the Social: An Introduction to Actor-Network-Theory.* New York: Oxford University Press, 2005.

Lieuwen, Edwin. *Mexican Militarism: The Political Rise and Fall of the Revolutionary Army, 1910–1940*. Albuquerque: New Mexico University Press, 1968.

Lorey, David E. *The University System and Economic Development in Mexico since 1910*. Stanford, CA: Stanford University Press, 1993.

Martín Hernández, Vicente. *La arquitectura doméstica de la ciudad de México (1890–1925)*. Mexico City: UNAM, 1981.

Matthews, Michael. *The Civilizing Machine: A Cultural History of Mexican Railroads, 1876–1910*. Lincoln: University of Nebraska Press, 2013.

McIvor, Arthur, and Ronald Johnston. *Miners' Lung: A History of Dust Disease in British Coal Mining*. London: Routledge, 2016.

Medina, Edin, Ivan da Costa Marques, and Christina Holmes. *Beyond Imported Magic: Essays on Science, Technology, and Society in Latin America*. Cambridge, MA: MIT Press, 2014.

Mendoza Vargas, Héctor. "El automóvil y los mapas en la integración del territorio mexicano, 1929–1962." *Investigaciones Geográficas, Boletín del Instituto de Geografía* (UNAM) 88 (2015): 91–108.

Mendoza, Vandari M. *Las patentes de invención mexicanas: Instituciones, actores y artefactos (1821–1911)*. Zamora: El Colegio de Michoacán, 2018.

Merchant, Luis Anaya. "Guerra, automóviles y carretera: La influencia norteamericana y el mercado automotriz mexicano en la 'reconstrucción' posrevolucionaria." *Boletín* (FAPECFT) 73, no. 1 (May–August 2013): 1–32.

Miranda Pacheco, Sergio. "El Frankenstein urbano: Ecólogos, urbanistas, e ingenieros frente a la crisis hidrológica de la ciudad de México a mitad del siglo XX." *Historia Ambiental, Latinoamericana y Caribeña* 10, no. 2 (2020): 162–202.

Montaño, Diana J. *Electrifying Mexico: Technology and the Transformation of a Modern City*. Austin: University of Texas Press, 2021.

Moya Gutiérrez, Arnaldo. *Arquitectura, historia y poder bajo el régimen de Porfirio Díaz: Ciudad de México, 1876–1911*. Mexico City: CONACULTA, 2012.

Muir-Wood, David. *Civil Engineering: A Very Short Introduction*. New York: Oxford University Press, 2012.

Neufeld, Stephen. *The Blood Contingent: The Military and the Making of Modern Mexico, 1876–1911*. Albuquerque: University of New Mexico Press, 2017.

Novo, Salvador. *Historia de la aviación en Mexico*. Mexico City: Compañía Mexicana de Aviación, 1974.

Núñez, Daniel Reséndiz. *La investigación en ingeniería: Consideraciones sobre su historia en México*. Mexico City: Instituto de Ingeniería, UNAM, 1979.

O'Rourke, Kathryn E. *Modern Architecture in Mexico City: History, Representation, and the Shaping of a Capital*. Pittsburgh: University of Pittsburgh Press, 2017.

Perló Cohen, Manuel. *El paradigma porfiriano: Historia del desagüe del Valle de México*. Mexico City: Porrúa, 1999.

Pezzoli, Keith. *Human Settlements and Planning for Ecological Sustainability: The Case of Mexico City*. Cambridge, MA: MIT Press, 2000.

Poniatowska, Elena. *Nothing, Nobody: The Voices of the Mexico City Earthquake.* Philadelphia: Temple University Press, 1995.

Pretel, David, and Lino Camprubí, eds. *Technology and Globalisation: Networks of Experts in World History.* London: Palgrave Macmillan, 2018.

Pretel, David, Ian Inkster, and Helge Wendt, eds. "History of Technology in Latin America." Special issue, *History of Technology* 34 (2019).

Prieto, Julie Irene. *The Mexican Expedition, 1916–1917.* Washington, DC: Center of Military History, United States Army, 2016.

Pritchard, Sara B., and Carl Zimring. *Technology and the Environment in History.* Baltimore: Johns Hopkins University Press, 2020.

Rath, Thomas. *The Myth of Demilitarization in Postrevolutionary Mexico, 1920–1960.* Chapel Hill: University of North Carolina Press, 2013.

Richmond, Douglas. *Venustiano Carranza's Nationalist Struggle, 1893–1920.* Lincoln: University of Nebraska Press, 1983.

Rodríguez Morales, Leopoldo. *El campo del constructor en el siglo XIX: De la certificación institucional a la esfera pública en la ciudad de México.* Mexico City: Instituto Nacional de Antropología e Historia, 2012.

Rogers, Thomas D. *The Deepest Wounds: A Labor and Environmental History of Sugar in Northeast Brazil.* Chapel Hill: University of North Carolina Press, 2010.

Rosental, Paul-André, ed. *Silicosis: A World History.* Baltimore: Johns Hopkins University Press, 2007.

Rosenzweig, Gabriel. "Los diplomáticos mexicanos durante la Revolución: Entre el desempleo y exilio." *Historia Mexicana* 61, no. 4 (April–June 2012): 1461–523.

Ruiz Romero, Manuel. *Aeropuertos: Historia de la construcción, operación y administración aeropuertaria en México.* Mexico City: Aeropuertos y Servicios Auxiliares, 2003.

Ruiz Romero, Manuel. *La aviación civil en México.* Mexico City: Universidad Nacional Autónoma de México, 1999.

Ruiz Romero, Manuel. *La aviación durante la Revolución Mexicana.* Mexico City: Soporte Aeronáutico, S.A., 1988.

Rúiz Romero, Manuel. *Grandes vuelos en la aviación Mexicana.* Mexico City: Grupo Editorial Aviación, S.A., 1986.

Saldaña, Juan José, ed. *Conocimiento y acción: Relaciones históricas de la ciencia, la tecnología y la sociedad en México.* Mexico City: Facultad de Filosofía y Letras-UNAM y Plaza y Valdés, 2013.

Saldaña, Juan José. *Las revoluciones políticas y la ciencia en México.* Mexico City: CONACYT, 2010.

Saldaña Solís, Marcela. "Luz y espacio: La modernidad en la obra constructiva de Emilio Dondé Preciat en la ciudad de México." *Boletín de Monumentos Históricos* 37 (May–August, 2016): 88–103.

Saldaña Solís, Marcela. "El Palacio Legislativo Federal y la participación de Emilio

Dondé, 1897–1902." *Boletín de Monumentos Histórico* 44 (September–December 2018): 168–79.

San Martín, Iván, ed. *Del Batallón al Compás: Cien años de aportaciones arquitectónicas de los ingenieros civiles (1821–1921)*. Mexico City: Universidad Nacional Autónoma de México, 2019.

Santiago, Myrna. *The Ecology of Oil: Environment, Labor, and the Mexican Revolution, 1900–1938*. Cambridge: Cambridge University Press, 2009.

Schwab, Stephen. "The Role of the Mexican Expeditionary Air Force in World War II: Late, Limited, but Symbolically Significant." *Journal of Military History* 66 no. 4 (2002): 1115–40.

Scott, James C. *Seeing Like a State: How Certain Schemes to Improve the Human Condition Have Failed*. New Haven, CT: Yale University Press, 1998.

Silva Contreras, Mónica. *Concreto armado, modernidad y arquitectura en México: El sistema Hennebique, 1901–1914*. Mexico City: Universidad Iberoamericana, 2016.

Smith, Michael M. "Carrancista Propaganda and the Print Media in the United States: An Overview of Institutions." *The Americas* 52, no. 2 (October 1995): 155–74.

Sociedad de Ex Alumnos de la Facultad de Ingeniería. *Ingenieros en la Independencia y en la Revolución*. Mexico City: Sociedad de Ex Alumnos de la Facultad de Ingeniería, 2010.

Soto Laveaga, Gabriela. *Jungle Laboratories: Mexican Peasants, National Prospects, and the Making of the Pill*. Durham, NC: Duke University Press, 2009.

Tanamachi Astro, Gerardo, and María De La Paz Ramos Lara. "La Escuela Nacional de Ingenieros y las Ciencias Físicas en los Albores del Siglo XX." *Revista Mexicana de Investigación Educativa* 20, no. 65 (June 2015): 557–80.

Tenorio-Trillo, Mauricio. *Mexico at the World's Fairs: Crafting a Modern Nation*. Berkeley: University of California Press, 1996.

Tinajero, Araceli, and J. Brian Freeman, eds. *Technology and Culture in Twentieth-Century Mexico*. Tuscaloosa: University of Alabama, 2013.

Tutino, John. *The Mexican Heartland: How Communities Shaped Capitalism, a Nation, and World History, 1500–2000*. Princeton, NJ: Princeton University Press, 2018.

Vergara, Angela. *Fighting Unemployment in Twentieth-Century Chile*. Pittsburgh: University of Pittsburgh Press, 2021.

Vergara, Germán. *Fueling Mexico: Energy and Environment, 1850–1950*. New York: Cambridge University Press, 2021.

Walker, Louise. *Waking from the Dream: Mexico's Middle Classes after 1968*. Stanford, CA: Stanford University Press, 2014.

Wolfe, Mikael D. *Watering the Revolution: An Environment and Technological History of Agrarian Reform in Mexico*. Durham, NC: Duke University Press, 2017.

CONTRIBUTORS

J. Justin Castro is a professor and chair of the Department of History at Arkansas State University. He is a specialist in the history of technology, engineering, and Mexico. He is the author of *Radio in Revolution: Wireless Technology and State Power in Mexico, 1897–1938* (University of Nebraska Press, 2016) and *Apostle of Progress: Modesto C. Rolland, Global Progressivism, and the Engineering of Revolutionary Mexico* (University of Nebraska Press, 2019).

Francisco Omar Escamilla-González is responsible for the Historical Archive of the Palace of Mining of UNAM. He specializes in mining technology issues in the eighteenth and nineteenth centuries; civil engineering in the late nineteenth century; the transfer of technical knowledge from Europe to Mexico, especially from Germany; and the circulation of technical and scientific books within Mexico. He is a member of the International Erbe Symposium Cultural Heritage in Geosciences, Mining, and Metallurgy.

James A. Garza is an associate professor and director of ethnic studies at the University of Nebraska. He is author of *The Imagined Underworld: Sex, Crime, and Vice in Porfirian Mexico* (University of Nebraska Press, 2008) and "Conquering the Environment and Surviving Natural Disasters," in *A Companion to Mexican History and Culture* (Wiley-Blackwell, 2011), edited by William H. Beezley. Currently Garza is at work on a transnational history that examines how in the late nineteenth century the Mexican government and the British engineering firm Pearson and Son helped transform the Basin of Mexico through the construction of the Gran Canal.

Rocio Gomez is assistant professor of Latin American history at Virginia Commonwealth University. She is the author of *Silver Veins, Dusty Lungs: Mining, Water, and Public Health in Zacatecas, 1835–1946* (University of Nebraska Press, 2020).

Lucero Morelos-Rodríguez has a doctorate in history and a research position at the Institute of Geology of UNAM, where she is responsible for the Historical Archive. She is a member of the Comarca Minera, Hidalgo, México, the UNESCO Global Geopark team, the International Commission on the History of Geological Sciences (INHIGEO), and Historiadores de las Ciencias y las Humanidades, A.C. Her research interests include the history of mining practices and the history of scientific institutions.

Jayson Maurice Porter is an environmental historian of Mexico and the Americas with a focus on oilseeds, agrochemicals, environmental justice, and ecological violence. He is an editorial board member of the North American Congress on Latin America (NACLA), a Voss Postdoctoral Fellow at the Institute at Brown for Environment and Society (IBES), and a Presidential Postdoctoral Fellow at the University of Maryland, College Park.

Juan José Saldaña has a PhD in the history of science from the University of Paris (Panthéon-Sorbonne). He is currently a professor of philosophy and history of science at the Universidad Nacional Autónoma de México (UNAM). He specializes in the social history of science and technology in Mexico and Latin America. He is the founder and director of *Quipu: Latin American Journal of History of Sciences and Technology* and secretary general of the International Union of History and Philosophy of Science/Division of History of Science. He has authored several monographs and articles on the history of science and technology in Latin America, and he was the editor of *Science in Latin America: A History* (University of Texas Press, 2006).

Pete Soland is a scholar of Latin American history specializing in modern Mexico. He earned his PhD in history from the University of Arizona in 2016. Currently, he is an assistant professor at Southeast Missouri State University in Cape Girardeau, Missouri. His forthcoming monograph, *Mexican Icarus: Aviation and the Modernization of Mexican Identity, 1928–*

1960 (University of Pittsburgh Press) examines the development of aviation in Mexico against the backdrop of the political, cultural, and economic processes that transformed the country in the decades following the revolution. He is currently researching the transnational history of nuclear programs in the Latin America.

Matthew Vitz is an associate professor of history at the University of California–San Diego. His research focuses on the intersection of environmental, political, and urban history. He is the author of *City on a Lake: Urban Political Ecology and the Growth of Mexico City* (Duke University Press, 2018).

INDEX